SpringerBriefs in Fire

Series Editor

James A. Milke

For further volumes:
http://www.springer.com/series/10476

Kathleen H. Almand

Structural Fire Resistance Experimental Research

Priority Needs of U.S. Industry

 Springer

Kathleen H. Almand
The Fire Protection Research Foundation
Quincy, MA
USA

ISSN 2193-6595 ISSN 2193-6609 (electronic)
ISBN 978-1-4614-8111-9 ISBN 978-1-4614-8112-6 (eBook)
DOI 10.1007/978-1-4614-8112-6
Springer New York Heidelberg Dordrecht London

Library of Congress Control Number: 2013942484

Printed on acid-free paper

Springer is part of Springer Science+Business Media (www.springer.com)

Foreword

This report provides a synthesis of stakeholder input to a prioritized agenda for research at the National Fire Research Laboratory (NFRL) at the National Institute of Standards and Technology (NIST) designed to accelerate the implementation of performance-based fire engineering for structures. It includes a comprehensive literature review of large-scale structural fire testing and a compilation of research needs from a variety of sources. It concludes with a prioritized set of research recommendations for the NIST facility.

The Foundation acknowledges the contributions of the following individuals and organizations to this project:

Dr. Luke Bisby, John Gales and Cristián Maluk, BRE Centre for Fire Safety Engineering, University of Edinburgh, Scotland, who conducted the literature review sections of this report;
and to:

Mike Engelhardt, University of Texas
Dan Howell, FM Global
Venkatesh Kodur, Michigan State University
Ehab Zalok, Carleton University
Greg Deierlein, Stanford University
Jim Milke University of Maryland
Greg Mueller, University of North Dakota
Maria Garlock, Princeton University
Marc Janssens, Southwest Research Institute
Mahmood Tabaddor, Underwriters Laboratories
Noureddine Benichou, National Research Council of Canada
Roberto Leone, Georgia Tech

who served as members of an ad hoc steering committee for the community input and workshop;
and to:

all those who participated in the community input meetings and the research needs workshop, with special thanks to workshop leaders Craig Beyler, Greg Deierlein, Venkatesh Kodur, Steve Szoke, Mike Engelhardt, and Charlie Carter.

The input and guidance from the NIST Engineering Laboratory sponsor representatives, Stephen Cauffman and John Gross, is gratefully acknowledged.

This report was prepared by the Fire Protection Research Foundation under award #60NANB10D181 from National Institute of Standards and Technology, U.S. Department of Commerce. The statements, findings, conclusions, and recommendations are those of the author and do not necessarily reflect the views of National Institute of Standards and Technology or the U.S. Department of Commerce.

Preface

This report provides a synthesis of stakeholder input to a prioritized agenda for research at the National Fire Research Laboratory (NFRL) at the National Institute of Standards and Technology (NIST) designed to accelerate the implementation of performance-based fire engineering for structures.

Those activities consisted of literature review, community input, and a one-day research needs workshop. The result was a large input of research topics which are synthesized in this report.

The NFRL presents a unique opportunity to explore a broad range of unanswered questions regarding the performance of real structures in the fire condition, and to inform performance-based design methods and standards in this field. Although input was sought from a broad range of sources in this study, the following major issues are of broad concern to the community and are recommended for consideration as priority research areas for the NFRL in support of its objective:

(1) Because of the unique nature of the facility, there is a priority need for a thorough benchmarking and validation process for the measurement systems used in the facility.

(2) Although the primary focus of the NFRL is structural behavior in fire, a priority for the design community is the interaction of real fire exposure and structural response—how one affects the other.

(3) A focus on large-scale experiments related to the many unanswered questions about composite floor system performance would have great practical import and a major impact on design methods.

(4) Material (structural and fire proofing) properties under load at the large scale are a high priority need for enhancing modeling of performance in fire.

(5) Understanding the embedded safety factors in our current prescriptive design methods is an important first step in moving toward a performance-based design system.

(6) There is a strong interest within the structural fire engineering research community in collaborating with NIST in undertaking synergistic research projects that take full advantage of the NFRL capabilities.

Contents

Abstract

This report provides a synthesis of stakeholder input to a prioritized agenda for research at the National Fire Research Laboratory (NFRL) at the National Institute of Standards and Technology (NIST) designed to accelerate the implementation of performance-based fire engineering for structures. It includes a comprehensive literature review of large-scale structural fire testing and a compilation of research needs from a variety of sources. It concludes with a prioritized set of research recommendations for the NIST facility.

Chapter 1
Introduction

In 1981, pioneering fire engineer Margaret Law presented a chapter at the ASCE Spring Convention in New York entitled "Designing fire safety for steel—recent work" (Law 1981). The visionary chapter presented a summary of novel work that she and colleagues at Arup Fire had completed to performance engineer the structural fire safety of innovative and architecturally exciting buildings—such as the Pompidou Center in Paris. Among the many topics covered in this chapter, Law (1981) stated a number of criticisms of the standard fire resistance test and proposed the way forward using knowledge-based analytical approaches. Paraphrasing, Law's key criticisms (directed predominantly at fire resistance testing of protected steel elements) were that:

1. the standard temperature–time curve is not representative of a real fire in a real building—indeed it is physically unrealistic and actually contradicts knowledge from fire dynamics;
2. the required duration of fire exposure in the standard test (or the time-equivalent exposure) is open to criticism on a number of grounds and should be revisited;
3. the loading and end conditions are not well defined—and clearly cannot represent the continuity, restraint, redistribution, and membrane actions in real buildings; and
4. the structural properties of the test specimen at room temperature are not well defined.

Some 30 years after this chapter was published the Structural Fire Engineering (SFE) community continues to base much of its guidance and regulatory compliance requirements on the standard fire resistance test (e.g. ASTM 2010; ISO 1999). A notable exception to this is in the use of so-called natural fires and performance based objectives in the design and analysis of steel-framed office buildings in Europe (Corus 2006). The fact remains; however, that the SFE community is only now beginning to truly wrestle with the full structure response of real structures (other than regular composite steel frames) in real fires. This is largely a consequence of the events of September 11, 2001, along with a few other notable

K. H. Almand, *Structural Fire Resistance Experimental Research*,
SpringerBriefs in Fire, DOI: 10.1007/978-1-4614-8112-6_1,
© Fire Protection Research Foundation 2012

structural failures (Beitel and Iwankiw 2008). It is also noteworthy that when structures do fail in fires (admittedly quite rarely) it is usually for reasons that would not (or could not) have been expected on the basis of standard testing.

In recent years, structural fire testing has experienced something of a renaissance. After more than a century with the standard fire resistance test being the predominant means to characterize the response of structural elements in fires, both the research and regulatory communities are now actively confronting the many inherent problems associated with using simplified single element tests on isolated structural members subjected to unrealistic temperature–time curves to demonstrate adequate structural performance in fires. As a consequence, a significant shift in testing philosophy to large-scale non-standard fire testing, using real rather than standard fires, is growing in momentum globally. A number of non-standard testing facilities have recently come on line or are nearing completion. Large-scale non-standard tests performed around the world during the past three decades have identified numerous shortcomings in our understanding of real building behavior during real fires; in most cases, these shortcomings could not have been observed through standard tests.

The National Institute of Standards and Technology (NIST)'s Engineering Laboratory is completing the construction of a structural fire resistance testing facility with the capability to test structural elements, systems, and their connections, which is unique in the current global fire research community. This capability, combined with other facilities at private and governmental laboratories, provides a resource to the structural fire engineering community to advance the validation of performance based fire engineering for structures.

The National Fire Research Laboratory (NFRL) will be built as an addition to the existing Large Fire Laboratory on NIST's Gaithersburg, Maryland Campus. It will include a 60 × 90 ft strong floor and a 60 ft long by 30 ft tall reaction wall. A 45 × 50 ft hood will capture the exhaust products from fires. The facility will have a capacity for testing structures up to two stories tall and two bays by three bays in plan. The facility has been designed to accommodate fires up to 20 MW in size.

The NSFRL will provide the capability to develop an experimental database on performance of large-scale structural connections, components, subassemblies, and systems under realistic fire and loading, to validate physics-based models to predict fire resistance performance of structures, and support the development of performance-based standards for fire resistance.

Chapter 2
Project Goal

To engage stakeholders in structural fire safety in a series of activities to develop a prioritized agenda for research at the new NSFRL facility designed to accelerate the implementation of performance based fire engineering for structures.

K. H. Almand, *Structural Fire Resistance Experimental Research*,
SpringerBriefs in Fire, DOI: 10.1007/978-1-4614-8112-6_2,
© Fire Protection Research Foundation 2012

Chapter 3
Review of Activities

The Foundation carried out the following activities in support of this goal:

1. An assessment of priorities derived from a review of worldwide large scale experimental facilities, research programs, and research needs assessments carried out in the past 20 years;

2. Development of a straw man assessment of research needs through dialogue and selected meetings with the structural fire engineering industry and research community to assess their priorities for large scale experimental research, including:

    (a) meeting with representatives from major structural and fire protection materials companies who participate in the American Society of Testing and Materials Fire Test Committee;

    (b) email dialogue with the Society of Fire Protection Engineers, American Institute of Steel Construction, American Concrete Institute, and American Society of Civil Engineers Technical Committees;

    (c) meeting with structural fire experts in lead engineering organizations who participate in the Council for Tall Buildings and Urban Habitat Fire and Safety Working Group;

    (d) meeting with the International Forum on Fire Research;

    (e) email dialogue with the SiFire Scientific Committee;

    (f) a teleconference with an Ad-Hoc Committee of U.S. based researchers with interests in structural fire engineering;

    (g) and other sources;

3. Development of a prioritized list of research needs through a one day meeting/workshop held in September of 2011 at the NIST facilities in Gaithersburg, MD.

K. H. Almand, *Structural Fire Resistance Experimental Research*,
SpringerBriefs in Fire, DOI: 10.1007/978-1-4614-8112-6_3,
© Fire Protection Research Foundation 2012

Chapter 4
Literature Review

4.1 Large-Scale Non-Standard Tests

The current review is focused on large-scale, non-standard structural fire resistance testing, with a particular emphasis on testing programs aimed at better understanding the full-structure response of real buildings in real fires. In general, the review addresses tests which have used real fires rather than testing furnaces to provide the thermal insult to the tested assembly. However, in some cases furnaces have been used to perform structural fire resistance tests which clearly fall outside the scope of standard testing procedures, and some of these have thus been included. In performing this review, more than 30 individual large scale tests have been identified since the early 1990s, although in some cases (as in the case of the Cardington steel frame tests, for instance) multiple tests were performed and are reported over a period of years in a single test structure. The following sections present brief overviews of the available tests, and identify common themes and conclusions wherever possible. The tests are categorized based on primary construction materials construction, and then chronologically.

4.1.1 Steel–Concrete Composite Construction

4.1.1.1 Stuttgart-Vaihingen University Fire Test (1985)

One of the earliest intentional experimental large-scale non-standard fire tests of the modern era was probably a test performed at the Stuttgart-Vaihingen University, Germany, in 1985. Very little information on this test is available in the literature. Indeed the authors of the current report can only locate a brief description of this test in the British Steel report on the Cardington tests (British Steel 1999).

It appears that the Stuttgart-Vaihingen University test was performed on the third floor of a four-story steel framed office building in which many different forms of

K. H. Almand, *Structural Fire Resistance Experimental Research*,
SpringerBriefs in Fire, DOI: 10.1007/978-1-4614-8112-6_4,
© Fire Protection Research Foundation 2012

steel and concrete composite elements had been used. The building incorporated water filled columns, partially encased columns, concrete filled columns, steel-composite beams, and various types of composite floor systems. The imposed fire was "natural" with a fire load consisting of wooden cribs. The structure was loaded by water-filled barrels and exposed to fire over about one-third of its floor area.

Despite fire temperatures in excess of 1,000 °C, the building maintained its structural integrity and experienced maximum beam deflections of only about 60 mm. The building was refurbished after the fire and subsequently occupied as office and laboratory space.

4.1.1.2 William Street Tests (Circa 1992)

In the early 1990s a series of four fire tests were performed by BHP, Australia's largest steel company, to investigate the structural fire performance of a specific steel-framed office building in Melbourne, Australia. The existing 41 story building (at 140 William Street) was undergoing renovation and refurbishment subsequent to an asbestos removal program. It was desired to show that, with the installation of a light hazard sprinkler system and a non-fire rated suspended ceiling, fire protection could be removed from the structural steelwork of the beams and the steel deck floors (British Steel 1999).

A test building was constructed to simulate a typical single story corner bay of the building; i.e. an isolated 12 × 12 m bay. The fuel load consisted of typical office furnishings amounting to about 53.5 kg/m^2. Gravity loads were applied using water tanks.

The first two tests were concerned primarily with the effectiveness of the proposed light hazard sprinkler system, and as such the fires were insignificant in terms of structural response. The third test examined the structural and thermal performance of the composite slab. In this test, the beams had 'partial fire protection' and the entire assembly was protected to some extent by a suspended ceiling system which remained largely in place during the fire. Peak gas phase temperatures were about 1,254 °C, but the composite slab supported the load. The fourth test used unprotected steel beams and composite slab, although still with a protective suspended ceiling, and had a peak gas phase temperature of 1,228 °C. The maximum beam temperature in this test was 632 °C, and the maximum beam deflection was 120 mm (no signs of impending collapse). However, it was found that 'most' of the 120 mm beam deflection was recovered on cooling. Three unloaded columns were also placed within the fire compartment during test 4 to investigate the performance of radiation shields; these were found to be extremely effective in these relatively low fire load conditions and kept column temperatures to about half of the value recorded for an unprotected column.

Taken together, the four William Street tests were used to 'demonstrate' that a light hazard sprinkler system was adequate to prevent deformation and collapse and that fire protection was not required on the beams or underside of the composite slab (British Steel 1999).

4.1.1.3 Collins Street Test (Circa 1994)

Shortly after the William Street tests, BHP performed a fire test in a steel-framed test building which was meant to simulate a section of a proposed multi-story building in Collins Street, also in Melbourne (British Steel 1999).

An 8.4×3.6 m compartment was loaded with 44–49 kg/m^2 of 'typical' office furniture. Peak gas phase temperatures of 1,163 °C were recorded, although the maximum steel beam temperature was only 470 °C. It is noteworthy that this test also included a suspended ceiling, which considerably reduced the temperatures measured in the structural elements, and that the structure had no imposed loading other than self-weight.

On the basis of this test, BHP successfully argued that no fire protection was required to the beams and the external steel columns. It has recently been noted (Vassart and Zhao 2011) that the Williams and Collins Street Fire tests have enabled—in conjunction with risk assessments and the use of sprinkler systems with a 'sufficient' (undefined) level of reliability—the use of unprotected beams in six multi-story office buildings between 12 and 41 stories in Australia up to 1999.

4.1.1.4 Cardington Steel Building Tests (1996)

During 1996, a number of large-scale non-standard structural fire tests were performed in an eight story composite steel-framed test building constructed at the Cardington test site of the UK Building Research Establishment. In total, seven distinct fire tests are reported in the literature, with the definitive reference for the first six of these being from the aforementioned research report by British Steel (1999). The building was 21×45 m, was three bays by five bays, and had a total height of 33 m. All beams were designed as simply supported, acting compositely with a 130 mm floor slab on steel decking. Beam-to-beam connections were made using fin-plate connections and beam-to-column connections using flexible end plates. Sandbags were used to simulate gravity loads for a typical office occupancy. A plan view of the test structure is given in Fig. 4.1, which shows the locations and sizes of the various fire tests performed.

Test 1 studied the behavior of a single 9 m long, internal restrained secondary beam (along with the surrounding floor slab) in an edge bay of the building, which was heated inside a custom built 8×3 m gas-fired furnace (refer to Fig. 4.1). The connections were outside the furnace and thus were not directly heated. The beam was heated relatively slowly (between 3 and 10 °C per minute) to a peak temperature approaching 900 °C while the temperatures and deflections of the structure were monitored. This test was intended to examine the effects of restraint on a heated beam from the surrounding cooler structure. The most notable conclusions from this test were that:

- the 'runaway' displacement which is typically observed in standard furnace tests of simply supported beams was not observed, despite maximum temperatures of 875 °C being observed in the beam's bottom flange;

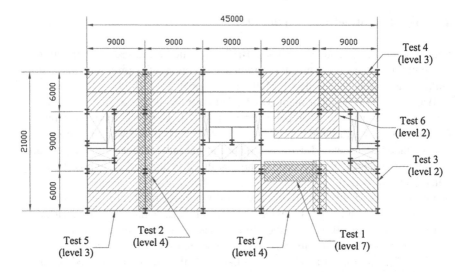

Fig. 4.1 Plan view of the Cardington Test frame and locations of the seven fire tests performed in this structure [reproduced after Vassart and Zhao (2011)]

- local buckling occurred at both ends of the beam just inside the furnace; and
- tensile failure of the beam-to-column connections occurred during the cooling phase of the fire.

Test 2 was used to study the behavior of a single story plane frame across the entire profile of the building. This included four columns and three primary beams, as shown in Fig. 4.1. The test was again performed inside a custom built gas fired furnace, in this case 21 × 2.5 m in plan. The top 800 mm of the columns and all beams and the underside of the composite slab were left unprotected, including the beam-column connections. The heating regime was similar to Cardington Test 1. The most notable conclusions from this test were that:

- the exposed portions of the columns buckled locally and squashed by about 180 mm when their temperature reached 670 °C;
- the column deformation caused a permanent deformation of 180 mm over all floors above the fire compartment (this is a highly significant result and indicates that fire protection of columns is of paramount importance); and
- it was noted that many bolts in the fin-plate connections between the primary and secondary beams (which had been heated over a length of only 1 m adjacent to the primary beams) had sheared due to thermal contraction of the secondary beam on cooling.

Test 3 was concerned with the behavior of the complete floor system, with a particular interest in membrane actions and alternative load paths. The test was performed in one corner of the 1st floor of the building in a 9 × 6 m corner-compartment (refer to Fig. 4.1), using a real fire load of 45 kg/m² of timber cribs. This resulted in a peak fire temperature of 1,071 °C. All columns, beam-to-column connections and edge beams were fire protected. The key conclusions from this test were that:

- a maximum vertical displacement of just less than span/20 occurred at the center of the secondary beam at a peak temperature of 954 °C (less than half of this deflection was recovered on cooling);
- the structure behaved 'well' and showed no signs of collapse; and
- buckling occurred near some of the beam-to-column connections although shear failure of the bolts did not occur.

Test 4 was also concerned with the behavior of a complete floor system, but also studied issues in compartmentation using steel stud partition walls. This test was conducted in a 54 m² corner compartment of the second floor of the building (refer to Fig. 4.1) using timber cribs with a load of 40 kg/m². All columns and connections were fire protected. The maximum recorded gas phase temperature was 1,051 °C after 102 min (the natural fire developed slowly in this case). Major conclusions were that:

- a large slab displacement of 269 mm occurred at the center of the compartment, which recovered to 160 mm on cooling;
- interactions between the fire exposed structure and the non fire exposed wind bracing above the test compartment were observed; these interactions are believed to have reduced beam displacements demonstrating the potential importance of capturing full-structure interactions during large scale structural fire tests; and
- No local buckling was observed and the connections showed no signs of high tensile forces on cooling.

Test 5 was a large compartment test intended to study the global behavior of the structure in a 21 × 18 m, side-compartment, spanning the entire width of the building on the second and third floors (Fig. 4.1). The compartment area was 340 m² and was uniformly fire loaded with wooden cribs at 40 kg/m². All of the steel beams, including the edge beams, were left unprotected; columns and connections were protected. The maximum recorded gas phase temperature was only 746 °C. Despite these relatively low temperatures, a number of potentially important conclusions were drawn, including that:

- A very large maximum slab displacement of 557 mm was recorded, and this recovered only to 481 mm on cooling;
- local buckling was observed near the beam-to-beam connections; and
- several the end-plate connections fractured down one side on cooling, and in one case the web detached the end-plate resulting in a total loss of shear capacity which caused large cracks in the composite slab above the connection but did not lead to collapse.

Test 6 was a demonstration test using a fire load of real office furniture to again study the global behavior of the structure in a large (18 × 9 m), open plan office corner-compartment of the building. In addition to the office furniture, wood and plastic cribs were added to give a fire load of 46 kg/m². Primary and secondary beams, including beam-to-beam connections, were exposed to the fire, while columns and beam-to-column connections were fire protected. A maximum gas phase temperature of 1,213 °C was recorded, with maximum temperatures of the unprotected steel in the range of 1,150 °C. The key conclusions from this test were that:

- maximum vertical deflections of 640 mm were observed and recovered to 540 mm on cooling;
- the structure showed no signs of failure, although the wind bracing in the floors above the test compartment likely offered alternative load paths (also the structure was able to rest on the block work walls forming the fire compartment in at least one location during the test); and
- the composite steel–concrete floor slab showed significant cracking around one of the columns during cooling; investigation revealed that the steel mesh reinforcement in the concrete slab had not been lapped correctly and simply butted together (this is clearly not desirable; however, it is possible that this could easily occur on a real building site).

Test 7, which is reported in detail by Wald et al. (2006), studied the global behavior of the structure in an 11 × 7 m side-compartment of the building and was concerned primarily with issues around tensile membrane action and the robustness of steel connections during fire. This test exposed two primary beams (unprotected), two columns (protected), and three secondary beams (unprotected) to fire (refer to Fig. 4.1) with a fire load consisting of 40 kg/m^2 of wood cribs. The maximum recorded gas phase temperature was 1,108 °C. Key conclusions were that:

- no collapse was observed despite maximum deflections up to 1,200 mm on heating which recovered by about 925 mm on cooling;
- buckling of the beams occurred adjacent to the joints during the heating phase resulting from restraint to thermal expansion provided by the surrounding structure; and
- as in Test 6, cracking of the concrete slab occurred at the column heads due to non-lapped steel mesh in the concrete slab.

Taken together, these seven tests demonstrated many important aspects of the full-structure response of composite steel-framed buildings during fire. In particular, they shed light on the secondary load carrying mechanisms, which can be activated during fire to prevent collapse, the potential importance of restraint to thermal expansion on heating and thermal contraction on cooling on localized buckling or connection failures, and the fact that full-structure response in fire is markedly different than that observed in standard fire resistance tests performed in furnaces. In the case of regular grid plan composite steel-framed buildings such as the one tested at Cardington, the fire resistance appears to be far greater than is normally assumed on the basis of furnace tests.

4.1.1.5 French Car Park Fire Tests (1998–2001)

A series of fire tests was conducted by the European Coal and Steel Council on an open-sided composite steel-framed car park structure. The objective of these tests was to show satisfactory structural fire performance without requiring fire protection to the structure. A single story structure 16×32 m in plan with a height of 3 m was constructed from unprotected steel columns, steel–concrete composite beams, and a steel–concrete composite slab with a depth of 120 mm. The structure was loaded with real cars, and the fire load in the most severe of the three tests consisted of three real cars burning simultaneously and exposing a 'significant area' of the floor to the flames of the fire (Vassart and Zhao 2011). This resulted in the steel beams above the fire being heated to more than 700 °C.

No collapse of the structure was observed during the tests, and maximum vertical deflections of only 150 mm were observed in the composite deck. These tests confirmed the positive influence of full structure response (particularly membrane actions) for composite steel car parks during fire. These tests have been used to develop 3D modeling tools and design tables for the fire-safe structural design of composite car parks which has been used in various fire safety engineering projects in France (Vassart and Zhao 2011).

4.1.1.6 Harbin Institute of Technology Tests (Circa 2007 and 2010)

Dong and Prasad (2009a) present the details of a large-scale experimental study to "understand the performance of structural frames under fire loading" and to "to serve as a database with which to check and validate numerical models." To do this, they constructed a two-dimensional, two-story, two-bay composite steel sway portal frame with fixed column base connections and subjected it to thermal loading by placing it inside a bespoke testing furnace while carrying sustained gravity loads.

Each bay of the frame spanned 3.6 m and each story was 2.8 m in height. The frame consisted of three vertical steel columns, linked together by four steel–concrete composite beams with composite action assured using shear studs. The depth of the reinforced concrete slab was 100 mm and the width of slab was 1,000 mm. Beam-to-column connections were bolted end-plate connections which were designed to transfer both moments and shear forces. Beams were unprotected but columns and beam-to-column connections were insulated. A unique testing furnace was custom designed to accommodate the frame while applying vertical loading with hydraulic jacks and while allowing for large displacements expected during testing. Full details of the testing facility are given by Dong and Prasad (2009a).

Three individual tests were conducted; these differed in the number of compartments that were heated by the furnace and in the relative location of the heated compartments. Tests included both heating and cooling phases using non-standard fires with peak gas phase temperatures in the range of 900 °C after between 60 and 115 min of fire exposure. Keeping in mind that this frame structure clearly cannot be considered fully representative of a real, three-dimensional structure, the key conclusions resulting from this study were that (Dong and Prasad 2009a):

- none of the columns in any of the three tests showed signs of local buckling or plastic hinge formation;
- structural failure was observed in the beam-to-column connections, the composite beam and the concrete slab. This resulted from contraction of the steel beams

on cooling and caused tensile connection failure, as has been observed by many authors studying the response to fire of composite steel structures;

- local buckling of the steel beams near the end of their heated spans was observed, again this has been seen in many prior tests by others;
- tensile cracking of the concrete slabs was observed near the span ends; and
- the deformation process and time to failure of a structure depend on the number and relative location of compartments that are heated in a structure (it should be noted that it is not entirely clear in this paper exactly how the authors have defined "failure").

Lv et al. (2011), from the same research group as the study above, have recently presented details of a study on "the performance of an edge beam in a full-scale three-story steel framed building subjected to fire." This paper is written in Chinese, however, and very few details of the study are currently available in English.

4.1.1.7 Mokrsko Fire Test (2008)

In September 2008, a large scale fire experiment was conducted in a purpose-built test structure by the Czech Technical University in Prague. The test was performed in Mokrsko, Czech Republic, and was designed to study the overall behavior of the test structure as well as its individual elements (Wald et al. 2010; Chlouba and Wald 2009). A problem in interpreting the results of this test is that six different wall structures and three types of flooring systems were all tested simultaneously, so that the impacts of localized failures of individual components are difficult to separate from the global response of the structure.

Wald (2011) states that the main objective of this test was to observe the temperatures of partially encased header plate connections, the behavior of castellated composite beams with the sinusoidal openings (called Angelina beams), and the behavior steel beams with thin corrugated webs. The test structure represented one floor of an assumed office building 12×18 m in plan with a height of 4 m. Wooden cribs providing a fire load of 35.5 kg/m^2 were used for the fire load and plastic bags filled by 'road-metal' were used to apply gravity loads representing a typical office occupancy.

Many interesting behaviors were observed during this test, however because the structure was such a mix of different components and systems it cannot be considered representative of a real structure and the results are of limited practical application.

4.1.1.8 FRACOF Fire Test (Circa 2008)

Vassart and Zhao (2011) present the results of two fire tests performed as part of the EU Commission-funded project Leonardo Da Vinci project on Fire Resistance Assessment of Partially Protected Composite Floors (FRACOF). The first of these tests was called the FRACOF test and was intended to investigate the applicability of the Cardington test conclusions for standard fires of longer duration (up to 120 min), for different construction details, and for the effect of higher gravity loading applied during fire. To investigate these issues, a single bay (representing a corner bay of a larger composite steel framed building was constructed. The building was one story high and 8.74 × 6.66 m in plan, with columns at the four corners, two primary beams spanning 6.6 m between the columns and four secondary beams spanning 8.74 m between primary beams. The 155 mm deep composite slab spanned 2.2 m between secondary beams. All four columns and the perimeter beams were fire protected, whereas the internal secondary beams and composite slab were left unprotected. Great care was taken around the perimeter of the structure to ensure good connection between the composite slab and the perimeter beams.

Loading was applied to the structure using sand bags equally distributed over the floor plate, and the entire bay was exposed to the ISO 834 (ISO 1999) standard fire for 120 min using a furnace (response was monitored also during the cooling phase). Full details of the results are avoided here, but the main observations/conclusions were that (Vassart and Zhao 2011):

- no collapse was observed for more than 120 min of standard fire exposure even with unprotected secondary steel beams, despite the failure of steel mesh reinforcement in the concrete slab;
- failure of the integrity and insulation criteria were due to the formation of a crack across the composite slab due to premature failure of the reinforcing steel mesh in the composite slab;
- proper overlapping of reinforcing steel mesh in the composite slab is essential to activate tensile membrane action and to ensure continuity of load transfer, particularly around columns;
- concrete cracking at the edge of the floor was limited and had no influence on the integrity and insulation performance of the floor (although it should be noted that the rotational restraint of the floor slab was probably marginal at the perimeter since no continuity of the floor plate was assured); and
- the floor behaved satisfactorily during the heating and cooling phases of the fire, as did all joints between steel members.

4.1.1.9 COSSFIRE Full Scale Fire Test (Circa 2008)

The second test described by Vassart and Zhao (2011) was called the COSSFIRE test and was very similar to the FRACOF test in scale but with a few minor modifications which allowed investigation of six different slab edge connections during standard fire exposure. Details of the test structure are omitted here, but it was essentially a composite steel framed structure with protected perimeter beams and columns and unprotected secondary beams and 135 mm deep steel–concrete composite slab. It is noteworthy that this test structure incorporated an unprotected secondary edge beam. The structure was loaded using sandbags and subjected to 120 min of the ISO 834 (ISO 1999) standard fire with response also monitored during the cooling phase. Key observations were that:

- the deflection of the floor was more than 500 mm after 120 min, although it behaved 'very well' and there was no sign of failure *in the central part of the floor*;
- the test was stopped due to large deflection and flexural failure of the unprotected secondary edge beam; however this failure did not lead to global collapse due to load redistribution from membrane effects;
- local buckling was observed in unprotected secondary beam connections, although all connections performed well during both heating and cooling phases;
- no failure of the edge connections between concrete slab and steel members was observed;
- cracking of concrete around columns, could have a negative impact on integrity criteria;
- no significant cracking of the concrete slab was observed in the central part of the floor, meaning that the reinforcing steel mesh behaved appropriately in membrane action up to 500 °C; and
- an edge detail of lapping steel reinforcing mesh in the concrete slab over shear studs on the edge beams was effective and should be applied in practice for this type of structure.

4.1.1.10 University of Ulster Fire Test (2010)

Vassart et al. (2010) present the results of a full-scale non-standard structural fire test performed by the University of Ulster (and collaborators) in Northern Ireland in February 2010. The fire test was on a composite floor system supported on long-span cellular steel beams. There was conducted to understand the development of tensile membrane action when the unprotected steel beams in the central part of a structural bay are cellular web, rather than solid web, beams.

The structure represented a single bay 15 × 9 m in plan with a height of 3 m. The surrounding walls were not fixed to the composite floor at the top to allow vertical movement of the floor. All the columns and perimeter beams were fire protected. The composite deck slab had a total depth of 120 mm and was fixed to edge beams by lapping the steel reinforcing mesh over shear studs welded to the perimeter beams. No external horizontal restraint was provided. A natural fire was imposed using wooden cribs at a load of 700 MJ/m^2, resulting in a gas phase temperature which peaked at 1,000 °C and lasted in excess of 90 min.

The main conclusions from this study are not yet available, however Vassart et al. (2010) state that the test was 'very successful' in that the structure performed 'as predicted'. They also state that using cellular secondary beams to support a composite slab (despite undesirable web-post buckling failures which were observed during the test) does not jeopardize the tensile membrane action that develops in such a slab in a fire situation.

4.1.1.11 TU Munich Fire Tests (2010)

Stadler et al. (2010) have briefly reported results of large-scale non-standard tests on a steel–concrete composite slab intended to enable structural fire designers in Germany to use membrane action for design of composite beam-slab systems in fire. The main objective of the project was to understand the behavior of intermediate beams between two composite slab panels. To do this, two fire tests were performed on composite slab panels 5×12.5 m in plan with unprotected secondary beams in different directions and configurations. All edge beams were protected with intumescent coating. The orientation of the secondary beams, the flooring system and the intumescent coating system were varied. One test used a lattice girder precast concrete slab and the second test used a profiled steel sheeting composite slab. The structure was exposed to a natural fire using wooden cribs to obtain gas phase temperatures which exceeded 900 °C during both tests. Very little information is currently available on the results from these tests, but it is expected that additional information will be available in the near future.

4.1.1.12 Veseli Fire Tests (2011)

Wald et al. (2011) present the results of two demonstration fire tests performed by the Czech Technical University in Prague in Veseli, Czech Republic during 2011 to investigate the 'design of joints to composite columns for improved fire robustness.' The tests were performed in a rather unusual and complicated, purpose-built test structure designed to represent a section of an office building constructed from steel–concrete composite beams with a steel deck slab and concrete-filled structural hollow section columns.

The two-story building was 10.4×13.4 m in plan with a height of 9 m and incorporated a number of innovative construction techniques, including steel fiber reinforced concrete slabs, concrete-filled structural hollow section columns, various external cladding systems, partial fire protection in various parts of the structure, and a number of different beam-to-column connection details. A natural fire exposure was imposed, consisting of non-uniformly placed timber cribs with a fuel load of 174 MJ/m^2 in the central region of the structure. As of 2011, detailed experimental observations and conclusions from these tests are not available in the literature.

4.1.2 Fire Tests on Portal Frames and Sub-Frame Assemblies

A number of medium and large-scale non-standard fire tests on steel or steel–concrete composite portal frame structures are available in the literature. These are not directly relevant to the current review since they do not simulate all of the requisite interactions that might play roles in real structures, although it is instructive to be aware of the work that has been performed.

Zhao and Shen (1999) experimentally studied the deformation behavior of unprotected two-dimensional, single-story, single span steel portal frames exposed to real fire conditions. Three tests were performed on steel frames with welded beam-to-column moment connections under different load levels and heating regimes within a custom-built gas furnace. Loads were applied only to the column heads, with no applied loads on the top beams of the portal frames.

Wong et al. (1999) performed an experimental study, apparently the only of its kind, investigating the fire performance of steel portal frame buildings. These tests involved a scaled portal frame building constructed from four portal frames with a span of 6 m at 1.5 m centers and with a rafter pitch of 15°. The building was clad with profiled steel sheeting. The base connection condition of the column was varied (pinned or fixed) amongst three separate tests. Pool fires were used to expose single portals to localized heating, with the columns remaining relatively cool, and the response and failure modes of the structure were observed.

Santiago et al. (2008) present details of a unique testing facility and a series of tests on six steel 'sub-frames' under a 'natural fire' at the University of Coimbra, Portugal, using a custom designed array of gas burners. The tests were intended to provide insight into the influence of various connection types on the behavior of steel sub-structures in fire. The test structures consisted of two thermally insulated columns and an unprotected steel beam with a 5.70 m span supporting a lightweight composite concrete slab. Pinned supports were imposed at the top and bottom of the columns. Six tests were performed, varying the beam-to-column connection configuration. These tests demonstrated the already well-known appearance of large tensile forces and reversal of bending moment during the cooling phase of a fire, as well as the fact that these forces may result in failure of the joint, as noted in many previous studies.

In addition to the work discussed previously on two-story, two-bay composite steel portal frames, Dong and Prasad (2009b) have also presented results from furnace tests on two full-scale composite steel frames with a height of 2.8 m and a span of 3.6 m. In one frame, both the beam-to-column connections and the columns themselves were protected, while in the other test only the beam-to-column connections were protected. The frames were subjected to both heating and cooling phases using a specially designed furnace. The results indicated that "the fire resistance of a composite beam is significantly better than that of a steel column," and that "the fire resistance rating of frames constructed with slim floor slabs is at least as good as that of frames with conventional floor slab construction."

Han et al. (2010) have recently presented results from a testing program on six planar portal frame structures with a span of 2.4 m and a height of 1.456 m in which the columns consisted of concrete-filled steel hollow structural sections and the beams consisted of reinforced concrete T-sections. These tests were intended to reproduce multi-story composite construction systems used in China. Test parameters included the shape of the columns (circular or square), the level of axial load in the columns, the load level in the beams, and the beam-column stiffness ratio. Only the beam was directly heated during these tests, following the ISO 834 standard fire. While omitting the details of the study and its conclusions, the test showed—rather importantly—that two failure modes were observed: column failure, or beam failure. The fire resistances of the frame were generally *lower* than those of individual concrete-filled columns tested in isolation but higher than those of isolated reinforced concrete beams. This demonstrates that, in certain scenarios, the performance of a structural element in a real building is worse than its performance would be in a furnace test.

4.1.3 Testing on Steel Connections in Fire

In addition to the large-scale non-standard fire tests presented above, a number of testing programs have been undertaken in the past two decades to investigate the performance of connections in steel structures subject to both standard and non-standard heating scenarios. Full details of these are not included here, but the interested reader could refer to Yu et al. (2009, 2011) for information on fire testing of steel bolted connections, Ding and Wang (2007) for information on fire testing of connections between steel beams and concrete-filled hollow structural sections, Chung et al. (2010) for tests on steel beam-to-column moment resisting connections, and Yuan et al. (2011) for information on fire testing of connections involving steel–concrete composite beams including the influence of the concrete deck slab.

4.1.4 Reinforced and Prestressed Concrete

Comparatively fewer large-scale, non-standard structural fire tests have been performed on reinforced or pre-stressed concrete structures, mostly because—in the absence of explosive cover spalling—they tend to perform very well in standard furnace tests as compared with unprotected steel structures. Nonetheless, some relevant testing is available in the literature, and it is described in the following sections.

4.1.4.1 'Non-Standard' Testing in Furnaces

While concrete structural elements (beams, slabs, columns, walls, etc.) all tend to perform well in standard fire resistance tests performed in furnaces, it is interesting to note that the concrete industry originally led the structural fire resistance community in performing what must be considered as 'unusual' furnace tests on concrete structural 'assemblies' rather than focusing immediately on standard tests. For instance, early testing on two way post-tensioned concrete slabs and beam-slab assemblies performed in the United States used combinations of beams and slabs to study the two-dimensional response of these structures (Troxell 1959).

A wealth of fire test data is available from these early tests; however, the focus in the current report is on testing performed since about 1985 so these 'historical' tests are not included here.

More recently, a number of authors have presented the results of fire tests on concrete elements or assemblies using modified standard furnaces to study specific structural response issues or specific types of structures which cannot be investigated using a standard approach. A non-exhaustive list of notable examples of this approach includes:

- Van Herberghen and Van Damme (1983), who used a modified standard floor furnace to study the fire resistance of post-tensioned continuous flat floor slabs with unbonded tendons;
- Kordina (1997), who used a modified floor furnace to investigate the punching shear behavior of reinforced concrete flat slabs in fire conditions;
- Kelly and Purkiss (2008), who used an oversized floor furnace to study the fire resistance of simply supported long-span post-tensioned concrete slabs; and
- Li-Tang et al. (2008), who studied the structural fire behavior of model-scale three-span continuous unbonded post-tensioned concrete slab strips in a custom built furnace subject to the standard fire;
- Zheng et al. (2010), who performed a series of tests on two-span continuous post-tensioned concrete slabs in a furnace with a central support built inside the heating chamber.
- Annerel et al. (2011), who used a modified standard floor furnace to perform punching shear tests on concrete slabs subject to the ISO 834 standard fire.

Many other examples are available in the literature, but these are not particularly relevant for the current review, which is more concerned with 'real', rather than standard, fire exposures.

4.1.4.2 Cardington Concrete Building Test (2001)

To the knowledge of the authors, only one large-scale natural fire test of an actual multi-story concrete building has ever been performed. Bailey (2002) presents the results of a natural fire test on a full-scale seven story cast in situ concrete building. The building was constructed as a demonstration building and to develop best practice guidance for different modern concrete technologies, and incorporated different concrete mixes, and construction techniques. The building was three bays by four bays and 22.5×30 m in plan. It had two cores which incorporated cross bracing for lateral load support. The concrete slab was 250 mm thick and the columns were either 400 mm^2 (internal columns) or 400×250 mm (perimeter columns). A number of different reinforcement layouts were used throughout the building; however, the reinforcement in the first and second floor slabs was traditional loose bar, with hook-and-bob links for shear resistance around the columns (Bailey 2002).

The main aim of the fire test was to investigate the behavior of a full-scale concrete framed building subjected to a realistic compartment fire and applied static design load. This entailed (Bailey 2002):

- investigating how the building in its entirety resisted or accommodated large thermal expansions from the heated parts of the structure;
- identifying beneficial and detrimental modes of whole building behavior that cannot be observed through standard fire tests;
- investigating the overall effects of concrete cover spalling and to determination of its significance on the behavior of the whole building; and
- comparing test results and observations from large-scale fire tests with current methods of design.

The fire exposure of the building was within a fire compartment at an edge bay of the building with an area of 225 m^2 on the ground and first floor, with an overall height of 4.25 m. The compartment walls were structurally isolated from the concrete slab and columns to prevent them from influencing the structural response to the fire. One internal column was exposed to the fire and eight additional columns were partially exposed to the fire. The columns were made from high strength concrete (103 MPa compressive cube strength) so 2.7 kg/m^3 of polypropylene fibers were added to the concrete mix for the columns. The cover to all reinforcement was specified as 20 mm. The structure was loaded during testing using sand bags. The fire consisted of timber cribs creating a fire load of 40 kg/m^2.

The full details of the test and its results are not presented here; however, the following key observations and conclusions were made:

- A maximum gas phase temperature of 950 °C was recorded 25 min after ignition, after which point the instrumentation was lost (the temperature is likely to have continued rising).
- Temperatures reduced between 12 and 13 min due to explosive spalling of the soffit of the floor slab. The spalling was extensive and reduced the severity of the fire throughout the test.
- Vertical displacements towards the edge of the building were larger than the displacements near the center and showed no signs of a plateau.
- Spalling to the soffit of the slab was extensive and exposed the bottom reinforcing bars. This was explosive and due to high in-plane compressive stresses in the slab caused by restraint to thermal expansion and high pore water pressures.
- The slab remained stable and supported the load by compressive membrane action at small slab vertical displacement. It should be noted that compressive membrane action can only occur at small displacements, and thus if the slab's vertical displacements were greater or lateral restraint surrounding the heated slab were less then 'it is difficult to see how the slab could have supported the static load.'
- Lateral thermal expansion of the slab resulted in lateral displacement of the external columns (in one case a 67 mm residual displacement was noted), which has been identified as the cause of previous failures of concrete structures during fires.

4.1.4.3 BRE Hollow-Core Slab Fire Test (2007)

In March 2007, two large-scale non-standard structural fire tests were performed by the UK Building Research Establishment at Middlesbrough, UK, to study the performance of hollow core concrete slab on steel beam flooring systems (Bailey and Lennon 2008). These tests were performed in the wake of worrying results from tests (e.g. Van Acker 2003) and incidences of failures of hollow core slabs in real building fires in Europe (e.g. De Feijter and Breunese 2007). The tests were intended to determine whether tying together and grouting of hollow core slabs in a floor plate is sufficient to prevent premature shear failure, which had been witnessed in small scale tests, and; therefore, to provide practical design guidance without the need for expensive and perhaps unnecessary tying.

Both tests were on a hollow core floor plate supported on protected steelwork and were identical except for the connection details between the units and supporting steel beams, with the second test having a more robust detail to tie the units and beam together. The fire compartment 7.0×17.8 m in plan and had a height of 3.6 m. The unprotected hollow core slabs sat on protected steel beams around the compartment perimeter. Fifteen 1,200 mm wide $\times$ 200 mm deep standard hollow core slabs with a concrete compressive strength of 85 MPa and a moisture content of 2.8 % by mass at the time of testing were placed in a single row to form the roof of the fire compartment.

In Test 1, the hollow core slabs sat directly on the supporting steel beams and the joints between the units and the gaps around the columns and units were grouted. In Test 2, U-shaped steel bars were placed in the cores and looped around shear studs fixed to perimeter beams. The cores were then grouted, as were the ends of the slab, the gaps between the slabs, and the gaps between the units and the columns. The slabs were uniformly loaded with sandbags and exposed to a natural fire using 32.5 kg/m^2 wooden cribs positioned evenly around the compartment; it was desired to follow the ISO 834 fire curve for the first 60 min of the fire exposure. The primary conclusions from these tests were that (Bailey and Lennon 2008):

- properly designed and detailed hollow core floor systems have inherent fire resistance and behave very well when subjected to very severe fire scenarios (more 'severe' than the standard fire);
- the maximum average gas phase temperature was over 1,000 °C in both tests;
- the hollow core floor performed well during the cooling phase of the fire, and the applied load was supported for the full duration of the fire;
- the edge units fractured locally during the cooling phase of the fire but this did not lead to loss of overall load carrying capacity;
- there was no significant spalling of the units despite their high concrete compressive strength;
- different end restraint conditions did not affect the measured vertical displacement; however, restraint conditions in Test 2 kept the outer proportion of the edge slab in place when it fractured along its length; and

- there was evidence of a lateral compressive strip forming at the ends of the units caused by restraint to thermal expansion which would have enhanced the flexural, and possibly shear, capacity of the units.

4.1.4.4   Hong Kong Fire Test (2010)

A large-scale non-standard fire test on unloaded concrete columns was performed in a real building in Hong Kong in August 2010 (Wong and Ng 2011). The purpose of this test was to study the effects of water quenching on the fire performance of high strength concrete in a building fire. The tests were focused specifically on various grades of high strength concrete local to Hong Kong under fire conditions, the effects of spalling of various grades of concrete with or without propylene fibers or wire mesh, and the effects of passive protective coatings on enhancing fire resistance of concrete structures.

Forty unloaded concrete columns of various types were placed inside a 'converted concrete pump house' structure that had been converted into a fire testing chamber and were subjected to a very severe pool fire. The key conclusion of the research project was that in order to prevent spalling in high strength concrete columns, insulating materials or fire-resisting coating materials should be used to restrict the heat transferred into the concrete, or polypropylene fibers or wire mesh should be provided in column members.

4.1.4.5   CCAA-CESARE Fire Test (2010)

Cement Concrete and Aggregates Australia (CCAA 2010) reported on a single large-scale non-standard fire test performed simultaneously on both high strength concrete columns and post-tensioned concrete slabs. The purpose of the tests was to understand the fire performance of post-tensioned slabs and high strength concrete columns made from concrete using common Australian aggregates (with aggregate type being a known risk factor for explosive spalling fire). As such, the tests aimed to assess the magnitude and extent of spalling in a real fire and to provide guidance on measures to limit its effects.

Twelve columns and three post-tensioned slabs were tested in a single fire enclosure. The columns differed in terms of concrete compressive strength, aggregate type, and the use of polypropylene fibers in the concrete mix. The columns were pre-compressed using a tensioned internal steel bar which was anchored at the column end to simulate the compressive loading to be expected in a real multistory concrete building. The three post-tensioned slabs differed only in terms of aggregate type.

The fire test was accomplished by placing the columns inside a plasterboard enclosure with internal dimensions of 4.25 × 5.4 × 3.3 m high. The post-tensioned slabs were placed over the roof of the enclosure. The fire load consisted of

124 kg/m^2 of wooden cribs, resulting in gas phase temperatures which exceeded 1,000 °C after about 45 min. From this test, the authors (rather unsurprisingly) concluded that (CCAA 2010):

- the addition of fibers to the high strength mix had a dramatic effect in reducing the level of spalling (the authors suggest that a dosage rate of 1.2 kg/m^3 is appropriate);
- the placement of column ties at closer spacing did not reduce the level of spalling;
- the post-tensioned slab containing one particular aggregate type spalled badly and hence consideration should be given to incorporating polypropylene fibers for this type of construction; and
- one of the post-tensioned slabs exhibited no spalling; whereas, an identical slab spalled at both ends with no obvious explanation.

4.1.4.6 TU Vienna Fire Tests (2011)

Ring et al. (2011) present the results of four large-scale non-standard fire tests on a 'frame-like' concrete structure in order to investigate the redistribution of loading within reinforced concrete structures subjected to fire (with a stated interest in buried infrastructure, i.e., tunnels). The tests were explicitly designed to provide data for the development, assessment, and validation of numerical tools for predicting the structural fire response of concrete structures, particularly for underground structures.

The frames were triangular and tubular in shape and were constructed on an exterior soil slope (refer to the source publication for figures of the rather unusual test specimens) and were loaded during testing to simulate a soil overburden. Two of the frames were made from concrete incorporating polypropylene fibers and two without fibers. The fire load supplied to the inside of the triangular tubes, with the bottom of the tubes insulated, using oil-burners following a pre-specified temperature history that rose to 1,200 °C in 9 min and remained at 1,200 °C for three hours.

The results of these tests provide important data for validation of computational models and they again clearly demonstrate the benefits of incorporating polypropylene fibers to prevent explosive spalling of concrete in fire.

4.1.4.7 University of Edinburgh and Indian Institute of Technology Roorkie Tests (2011)

Sharma et al. (2012) reported on a test performed jointly (in India) by the University of Edinubrgh and the Indian Institute of Technology, Roorkie, on a two-story full-scale reinforced concrete frame subjected to simulated earthquake and fire loads. The objective was to generate useful information on the behavior of damaged and undamaged reinforced concrete structures subjected to fire.

Three tests are planned during this project, although only the first of these has yet been reported in the literature. The test structures, which are constructed on a custom designed outdoor foundation with an adjacent reaction wall, are reinforced concrete frames consisting of four columns spaced 3 m apart in plan supporting four beams and a concrete slab. The structures are monolithic reinforced concrete. The frames are first subjected to cyclic quasi-static lateral loading to simulate the damage expected under seismic loads, then subjected to fire under gravity loads only (applied using sandbags), and finally loaded to failure laterally.

The fire exposure of the test structure consisted of a compartment fire on the ground level. A $3 \times 3 \times 3$ m compartment was built inside the structure beneath the first floor concrete slab, and a 1 m^2 kerosene pool fire (continuously fed) was used for fuel. This resulted in peak gas phase temperatures in the range of 900–1,000 °C for more than one hour of fire exposure. The columns were exposed to the fire. Only limited results from the first test are available, these indicate that the test frame could withstand a pre-damage corresponding to a seismic event and subsequent one-hour fire exposure without collapse. No spalling was observed.

4.1.5 Timber-Framed Buildings

To the knowledge of the authors, only one large-scale non-standard structural fire test of a timber structure has ever been presented in the literature. Lennon et al. (2000) present results from a large-scale compartment fire test in a full-scale six-story timber frame building, again at the UK Building Research Establishment's Cardington Facility in September 1999. The purpose of the test was to evaluate and demonstrate the performance of medium-rise timber frame buildings subject to real fires.

The fire was provided by uniformly distributed timber cribs and was imposed within a single apartment on level three of the building. Flame spread was uninhibited and ventilation was arranged so as give the worst case fire severity. A key test objective was to evaluate the effectiveness of fire compartmentation in preventing fire spread from the flat of origin to adjoining flats and in maintaining the integrity of the means of escape and structural stability (Lennon et al. 2000). Full details of the test are omitted here, but the following general conclusions were drawn:

- The performance of a complete timber frame building subject to a real fire is at least equivalent to that obtained from standard fire tests on its individual elements.
- Timber frame construction can meet the functional requirements of the Building Regulations in terms of limiting internal fire spread and maintaining structural integrity.
- The standard of workmanship is of crucial importance in providing the necessary fire resistance performance, especially nailing of plasterboards, and in ensuring the correct location of cavity barriers and fire stopping.

A series of fire tests on a timber building has also been presented by Peng et al. (2011) with limited application to structural fire engineering.

4.2 Large-Scale Non-Standard Fire Testing Facilities

A number of dedicated fire testing facilities are available globally that can be used to conduct large-scale non-standard fire tests, and a number of new facilities have recently come on line, are under construction, or are in the planning stages; these can be divided into (1) live fire test facilities, (2) modified fire testing furnaces, and (3) facilities under construction or in the planning stages.

4.2.1 Live Fire Test Facilities

A number of 'live fire' testing facilities are available globally. In addition to the numerous outdoor test sites which have been used to perform large-scale non-standard structural fire tests in recent years (e.g., British Steel 1999; Vassart and Zhao 2011; Chlouba et al. 2009; Wald et al. 2010; Vassart et al. 2010; Stadler et al. 2010; Wald et al. 2011; Ring et al. 2011; Sharma et al. 2012, all described previously), many government and private sector fire testing laboratories have large burn halls which can be used to perform fire tests of real structures using natural fires. Examples include (in no particular order):

- the University of Victoria's CESARE fire testing facility in Australia, which houses a 15 MW oxygen calorimeter for measurement of combustion products during large experimental fires (Victoria University 2011);
- the Building Research Establishment's (BRE) large burn hall, which has a 10 MW calorimeter and a smoke management used for determining heat release rates, in the United Kingdom (BRE 2011);
- the National Research Council of Canada's (NRCC) burn hall in Almont, Canada, which is 55 × 30 m in plan with a ceiling height of 12.5 m (NRCC 2011);
- the FM Global Fire Technology Laboratory, USA, houses a large burn hall with a footprint of 3,120 m^2 which includes a 20 MW fire products collector (FM Global 2009);
- Southwest Research Institute (SwRI), USA, houses a "state-of-the-art full-scale furnace facility and a full-scale indoor test facility" (SwRI 2011);
- The Tianjin Fire Research Institute apparently has a very large burn hall at its South River Test Site, China, although very few details regarding this facility and its capabilities are not available in English (TFRI 2011); and
- The SINTEF research laboratory, Norway, has a 590 m^2 (36 × 16.5 m) burn hall with a ceiling height between 22 and 28 m, designed for all types of large-scale fire tests and to withstand the heat and smoke load of a 12 m^2 gasoline pool fire with 18 m high flames (SINTEF 2011).

These live fire testing facilities typically do not have the capability of performing tests with variable or lateral structural loading, and many do not offer calorimetry to determine actual heat release rates nor smoke analysis for density and toxicity. The largest such live fire facility, the Building Research Establishment's Cardington Test Hall (British Steel 1999), is no longer available for use by the structural fire engineering community.

In addition to these live fire research facilities, many national and private sector fire research organizations have one or more standard structural testing furnaces which can be (and in many cases have been) modified to perform non-standard tests of structural elements. Such furnaces of various types, sizes, and configurations are available at most of the laboratories listed above.

4.2.2 Modified Furnaces

A number of custom designed furnace testing facilities are also available globally, which can be used to study various aspects of the response of structures and structural components to fire in a variety of non-standard ways. These may be based on existing standard fire testing furnaces, or may be custom designed from scratch and based either on gas-fired burners or on electrical heating. A few notable examples are given below.

The Fire Safety Engineering Research and Technology Centre at the University of Ulster, Northern Ireland (FireSERT), has a reconfigurable standard floor furnace, which has been modified to permit testing of specimens larger than would be typical. The furnace can also test beam-wall assemblies.

The University Coimbra, Protugal, has recently developed a special 'natural fire' testing rig which is capable of testing steel sub-frame assemblies under exposure to gas burners with the ability to spatially and temporally vary the thermal insult to loaded test assemblies (Santiago et al. 2008). Tests performed using this facility are briefly described in Sect. 4.1.2.

An example of a system, which is based on a pre-existing standard testing furnace, is the column testing facility at the Federal Institute for Materials Research and Testing (BAM), in Berlin, Germany. The BAM column testing facility is based on a standard column furnace but rather than simply applying a constant axial testing load during fire, it uses a 'hybrid' sub-structuring system where the entire column is tested inside the furnace and measured forces are read and target displacements are controlled via simulated modeling into the substructure in real time (Korzen 2010). The BAM furnace is capable of producing an ISO 834 (ISO 1999) fire on columns of up to 3.55 m in height. The furnace is fired by six oil burner and six electro-hydraulic control channels are used to actively control mechanical boundary conditions (two bending rotations each at top and bottom, axial displacement at the bottom, and horizontal displacement at the top). This testing methodology allows single elements to be tested in isolation while the rest of the structure is 'simulated' by actively controlling the boundary conditions during the test.

A similar approach has recently been taken at the National Research Council of Canada, Ottawa, Canada, where an existing standard column testing furnace has been upgraded to permit 'hybrid' testing of isolated structural elements by active control of applied loads during testing using coupled numerical analysis of the full structure in real time (Mostafaei 2011).

An example of a unique fire testing furnace based on electrical heating is the University of Sheffield's connection testing furnace (Yu 2011). This relatively small-scale facility was custom designed to allow testing bolted steel connections at elevated temperature under the appropriate combinations of moment, shear, thrust, and rotation that would be expected in a real building in a fire. Heating is accomplished using electrical radiant heaters, and loading is accomplished using a complicated mechanism of restraining frames and loading jacks; it allows detailed, rational studies of the response of bolted steel connections under a wide range of possible structural and thermal actions.

Probably the best example of a structural testing facility which is currently available for large-scale non-standard structural fire testing is the CERIB-Prométhée testing facility, in Epernon, France, which has been operational since 2008 (Robert et al. 2009). The facility was developed to expand existing standard tests, to incorporate boundary conditions which would occur due to restraint and interaction between hot and cold zones in a real fire in a real building, and to provide validation data for thermo-mechanical simulations of real structures during fire.

The facility essentially consists of a gas-fired furnace with integrated mechanical loading frames that can reproduce interactions between the parts of the structure under testing and those that are unexposed. The furnace measures $6 \times 4 \times 2.5$ m (adjustable to 4.1 m), although a specimen length of 10 m can also be installed due to the modular furnace construction. Restraint and structural interactions are simulated via multi directional loading using an array of some 29 hydraulic jacks with loading capacity up to 3 MN. These jacks can be actively controlled by coupling to numerical analysis of the full structure in real time. The facility is capable of testing floors, connections, tunnel linings, beams, slabs, and walls. The furnace consists of 16 gas burners and is capable of simulating various standard fires, including ISO 834 (ISO 1999) and hydrocarbon curves (CEN 2002).

4.2.3 Facilities Under Construction

In addition to the facilities noted above, a number of unique fire testing facilities are currently under construction. Notable examples are the construction of a new 9 m tall testing furnace at the Centre Scientifique et Technique du Bâtiment (CSTB), Paris, France, and the current large-scale extension to the National Fire Research Laboratory at the US National Institute of Standards and Technology (NIST).

The CSTB facility under construction, called 'Vulcain', is essentially a particularly large modularized and reconfigurable standard testing furnace which will be able to test walls of 3 m width and 9 m in height, long span floors with a weight of up to 30 tonnes, and various combinations of walls, columns, beams, and slabs in two and three dimensions (CSTB 2011).

NIST's extension to the National Fire Research Laboratory (NFRL) will create a unique live fire testing facility which combines most of the best aspects of available testing facilities elsewhere (NIST 2011). The facility will combine the capability to test large-scale multiple bay, multi-story structures, subject to real fires with real fuel loads, while applying controlled loads both vertically and laterally and providing data on heat release rates and gas analysis. The test area will consist of a 60 × 90 ft strong floor with an adjacent 30 ft high × 60 ft wide concrete strong wall. The strong wall will act to stabilize a test specimen to prevent uncontrolled failure, provide lateral restraint, or to laterally load a structure to simulate earthquake damage. A 45 × 50 ft hood, centered above the strong floor, will capture and remove smoke and hot gases and provide heat release rate calorimetry data up to fires 20 MW in size (NIST 2011).

4.3 Prior Reviews of Large Scale Non Standard Research

4.3.1 Bietel and Iwankiw (2008): NIST GCR 02-843-1 (Revision)

A report by Bietel and Iwankiw, which was originally commissioned by The National Institute for Standards and Technology and authored in 2002, was republished in 2008 with minor revision. The purpose of the report was "to analyze the needs and existing capabilities for full-scale fire resistance testing of structural connections," and consisted of a survey of historical information on fire occurrences in multi-story buildings that resulted in structural collapse, a survey of private and public facilities capable of testing the structural integrity of building elements under fire conditions, and an assessment of the needs for additional testing and/or experimental facilities to allow the performance of structural assemblies and fire resistance materials to be predicted under extreme fire conditions within actual buildings and options for meeting those needs (Bietel and Iwankiw 2008).

The report contains descriptions of 22 cases where multi-story buildings experienced fire-induced collapses between 1970 and 2002 with approximately equal distribution between steel, concrete, and masonry buildings. The majority of fires were in office/commercial buildings. While the specific details of these collapses are omitted here, it is fair to say that in most cases the critical failure modes observed could not have been predicted on the basis of standard furnace testing, that structural interactions and connections played important roles in all cases, and that in several cases the collapses occurred during the cooling phase of the fire. Based on this review of fire-induced collapses in real buildings, the authors note that "connections are generally recognized as the critical link in the collapse vulnerability of all structural framing systems, whether or not fire is involved." Bietel and Iwankiw (2008) also present a review of high-rise building fires without collapses, but with major structural damage, which leads them to suggest that further work on the structural fire response of the entire building framing should be conducted to develop a better understanding of structural fire safety, both in steel and concrete construction. Finally, brief comments are included on fires following earthquakes and fires in low-rise construction.

Bietel and Iwankiw (2008) subsequently present a global survey of structural fire resistance testing capability, by surveying fire research laboratories around the world as to their capabilities with respect to both vertical and horizontal structural elements. The focus of this survey is on structural fire testing capabilities involving *furnace testing*, as opposed to using natural fires. Based on the responses, the authors conclude that many laboratories are able to perform standard furnace fire resistance tests of various sizes, types of fire exposure (heating but not necessarily controlled cooling), loading, and measurements. They note that several unusually large furnaces exist and that these could be used to evaluate structural connections or combinations of building elements, but that *no single laboratory is currently able to test large-scale structural assemblies under the full range of applicable loading and fire exposure conditions.*

In terms of a "needs assessment" for structural fire resistance testing, Bietel and Iwankiw (2008) begin by highlighting the limitations of standard fire resistance testing, all of which have been previously noted and have been very well known and acknowledged since the late 1970s (or earlier). The major research needs noted include (paraphrasing):

- developing a better understanding of the interactions between structural elements and the thermally-induced loadings in a building during fire;
- understanding opportunities for load redistribution and alternative load paths in buildings to prevent global failure;
- demonstrating the performance and robustness of connections in fire;
- investigating the impacts of multiple floor fires and heating of elements from both sides;
- performing fire tests at realistic scale, particularly large members with long spans;
- developing better instrumentation for use during fire tests;
- understanding and demonstrating the reliability of structural fire protection materials; and
- development of a unique testing facility to accommodate the required size, appropriate loading and the fire exposures needed for longer, wider or taller members.

4.3.2 Wald (2011)

As part of the European COST Action TU0904: Integrated Fire Engineering and Response Working Group, a summary of available large scale structural fire tests conducted globally was presented by Frantisek Wald at a COST Action TU0904 Working Group meeting in April 2011 (Wald 2011). His paper presents a summary, with source references and brief descriptions, of a number of large-scale non-standard furnace tests, all of which are described previously.

4.3.3 Vassart and Zhao (2011)

A publication was produced by Vassart and Zhao (2011) as a result of the EU Commission-funded project Leonardo Da Vinci project on Fire Resistance Assessment of Partially Protected Composite Floors (FRACOF). This engineering background document includes detailed summaries of a number of large-scale non-standard structural fire resistance tests performed in Europe, including the Cardington tests and tests on open, steel-framed car parks performed in France. It also presents evidence from the accidental real fires at Broadgate and Churchill Plaza (discussed previously), and from small-scale fire resistance tests performed at the University of Manchester, UK (Bailey and Toh 2006).

Chapter 5
Research Needs Identified Through Literature Review

5.1 Fire Exposure

As noted by authors going back to at least 1981, the standard temperature–time curve is not representative of a real fire in a real building. In order to truly understand the response of real buildings in real fires, tests of structures and structural elements are required under credible worst case natural fire exposures. Depending on the type of structure and the occupancy under consideration, this may require experimental consideration of localized, compartmentalized, horizontally and/or vertically travelling, smoldering, or hydrocarbon fires, all of which have the potential to introduce structural actions or interactions which are not captured by the standard fires. There appears to be no facility in the world which is capable of producing (and reproducing) truly natural fire exposures on structures of realistic scale and construction.

5.2 Structural Interactions and Asymmetry

The available test data from large-scale non-standard fire tests, while expensive, still cover only a small fraction of the possible structural configurations that are represented within the current global building stock let alone the highly optimized and sustainable buildings of the future. With a few notable exceptions, the majority of structural fire tests conducted to date, whether standard or non-standard, have studied regular, symmetric, highly idealized structures. Modern structures increasingly make use of irregular floor plates with varying span lengths, bay sizes, construction types, etc. The possible influence of irregular floor plans and complex building forms needs to be investigated, both experimentally and numerically, if performance-based structural fire engineering of both conventional and modern buildings is to be performed with confidence.

K. H. Almand, *Structural Fire Resistance Experimental Research*,
SpringerBriefs in Fire, DOI: 10.1007/978-1-4614-8112-6_5,
© Fire Protection Research Foundation 2012

5.3 Failure Localizations

As already noted, when real structures fail in fires it is rarely for the reasons that might be expected on the basis of standard fire resistance testing. In many cases, failure is precipitated by some form of localized failure or distress, such as discrete cracking in concrete, rupture of tensile steel reinforcement, failure of a connection, local buckling of structural steelwork, shear (punching) failure of a concrete slab, etc. Unfortunately, the only way to observe and understand such failure localizations, which depend in virtually all cases on three dimensional interactions between elements of a structure during both heating *and* cooling, is to perform large-scale non-standard structural fire testing of real buildings. Only once the possible failure modes are known can they be rationally incorporated into computational models for full structure response.

5.4 Compartmentation and Fire Spread

To date, most large-scale structural fire testing has focused on prevention of structural collapse during fire, and relatively little attention has been paid to preservation of compartmentation under large deformations in real structures during fire, particularly given the large floor plate deflections and wide discrete cracking which are sometimes observed in large-scale fire tests on composite slabs. The impacts of vertical and lateral deformations of structural frames on fire stopping and both horizontal and vertical compartmentation should be studied in order to preserve life safety in buildings, which are now becoming ever more reliant on so-called defend-in-place life-safety strategies (for instance in highrise buildings where fire safety strategies are often fundamentally based on the assumption that a fire will be confined to the floor of origin and phased evacuation is put in place).

Furthermore, given that many structural fire engineers have serious concerns about the quality of installed fire stopping between floors in multi-story buildings, large-scale non standard fire tests should perhaps be considered in which vertical fire spread is simulated using natural fires, to evaluate the structural impacts of credible worst case fires burning simultaneously on more than one floor of a structure.

5.5 Detailing and Construction Errors

Taken together, the tests described in this report highlight a number of important construction details and potential construction errors which may appear inconsequential to a building contractor but which may have a profound impact on the structural fire response and integrity of a building during fire. Examples of this include integrity of fire stopping during large deformations, lapping of steel reinforcing mesh, anchorage of steel reinforcing mesh over shear studs on perimeter beams, use of deformed versus smooth bars for reinforcement (potentially leading to strain localization and tensile failure of deformed steel bars during fire), proper anchorage and grouting of hollow core slabs, use of specific types of bolted steel connections to promote connection ductility and rotational capacity during fire, quality, uniformity, and robustness of structural fire protection materials (either passive or intumescent), and so on. Serious unknowns continue to surround many, if not all, of these issues, and there is a need for testing to support the development of best practice guidance which can be used to support quality assurance programs on construction sites of fire engineered buildings.

5.6 Cooling Phase Behavior and Residual Capacity

A number of localized structural failures or adverse structural responses of both steel connections and concrete flat plate slabs have been observed during the cooling phase of both real fires in real buildings (e.g., Firehouse.com 2004; Bamonte et al. 2009) and non-standard heating regimes in large-scale structural fire experiments (e.g., Bailey and Lennon 2008; British Steel 1999). Structural actions resulting from creep, localized and global plastic deformation, local buckling, and thermal contraction and restraint, all need to be better understood for all types of structures if designers are to realistically be expected to design for burnout of a fire compartment without collapse.

Furthermore, the residual structural capacity of fire damaged structures that have undergone large deformations is not well known, meaning that many fire-damaged structures will need to be demolished after a fire (e.g., New York Times 1997). This is particularly true for so-called fire-engineered composite steel frames, which explicitly rely on large deformation behaviors to mobilize tensile membrane action in fire which are necessary to support gravity loads (British Steel 1999).

5.7 Instrumentation and Measurement

Most researchers agree that much more complete data are required from both standard and non-standard structural fire tests. Better information on strains and displacements during testing would allow a more accurate understanding of response, and would provide additional data which are essential for high quality model development and validation. The need for new types of sensors, such as wireless sensors to be used during fire tests, has been noted previously (Kodur et al. 2011). However, the authors of the current review feel that what is really needed is a better understanding of what is being measured; i.e., *What should be measured in order to truly understand the global performance of the element of the structure being tested?*

5.8 Data for Model Calibration, Validation and Verification

Experimental data are essential for calibration, validation, and verification of both existing and emerging computational modeling techniques to simulate the response of structures and structural elements in fire. This requirement holds both at the material level and at the structural (i.e., system) level. As noted by Kodur et al. (2011), high-temperature constitutive material models are needed to generate reliable input data for structural models to better predict system response to fire and possible failure modes. Such data must be developed using an appropriate framework for understanding the stress-temperature–time-strain interrelationships at play in most engineering materials. An excellent framework for materials characterization at elevated temperature has been presented by Anderberg (1986), but the complexities shown in this framework are rarely explicitly acknowledged in design.

5.9 Structural Optimization and the Use of New Materials and Systems

Modern structures are highly optimized, increasingly with the use of sophisticated computer analysis, in an attempt to reduce the mass, cost, environmental impact, carbon emissions, and embodied energy in buildings (Terrasi 2007). Modern structures also increasingly make use of innovative materials, such as high strength, self consolidating concrete, fiber reinforced polymers (FRPs), structural adhesives, stainless steel, etc.; and innovative structural systems, such as unbonded post-tensioned flat plate concrete slabs, the response of which during fire is not well known in many cases. New materials and structural systems must be rationally understood before they can be applied with confidence in buildings; such an understanding demands large-scale non-standard fire testing, in particular because the standard furnace tests that were developed for conventional construction materials and systems are based on structural response and failure definitions which often are not applicable to the innovative ones (Bisby and Kodur 2007).

5.10 Connections

As noted previously, a range of studies have already been performed on connection performance in fire (largely for steel structures) (e.g., Ding and Wang 2007; Yu et al. 2009, 2011; Yuan et al. 2011). However, given the range of possible connection types, full-structure responses to fire, and failure modes, additional research is needed to better understand the full range of possible connections, to develop and validate computational modeling capabilities to predict connection response, and to suggest best practice guidance to steel fabricators on the types of connections which should be applied in practice to ensure structural robustness in fire. Proper details for the connection of precast concrete elements in buildings to ensure robust performance in fire is also required (Bailey and Lennon 2008). Important lessons can be learned on these issues by studying the literature and available design provisions on the seismic design of structural connections (FEMA 2000a, b; AISC 2005); it may be appropriate to develop similar provisions for structural robustness against fire.

5.11 Explosive Spalling of Concrete

Structural fire design of modern reinforced and pre-stressed concrete structures relies on the assumption that the concrete will not spall during fire. This assumption is based largely on data from large-scale standard fire tests of concrete elements tested in isolation in furnaces during the past 60–70 years. However, there is legitimate concern (Kelly and Purkiss 2008) that modern concrete structures, which incorporate concrete mixes with considerably higher concrete strengths, are more susceptible to spalling than was historically the case. While preliminary guidance on the means by which spalling can be addressed by designers is available in the structural Eurocodes (CEN 2004), research is badly needed to understand the respective roles of the various factors which are known to increase a concrete's propensity for spalling during fire (e.g., high strength, high stress, high moisture, low permeability, small amounts of bonded reinforcement, use of silica fume, rapid heating, etc.) (Arup 2005; Bailey and Khoury 2011), such that defensible preventative actions can be taken (for instance the requirement to add a certain amount of polypropylene fibers to the concrete mix). Interactions in real structures have the potential to significantly influence development of spalling in a fire, so large scale tests under natural fires are needed to truly understand propensity for, and consequences of, spalling in real buildings.

5.12 Timber Structures

The critical issue in the structural fire protection of timber structures is the integrity of fire proofing materials such as gypsum plasterboard (Lennon 2000). Additional research is required to better understand the factors leading to 'fall-off' of plasterboard during fire.

Chapter 6
Research Needs Identified Through Community Input

As a result of the community dialogue and meetings held over the course of this project, an initial framework for research needs was developed with input from the steering committee. Although not a robust description of research needs and priorities, this framework was the starting point for the subsequent research needs workshop. This framework, with selected examples identified is illustrated below.

6.1 Large Frame Studies

- Steel structure
- various floor systems, connection types, moment frames
- impact of varying fire location
- explore the interaction of fire spread with the structural system (for example, the collapse of barrier walls changing the fire compartmentation)
- system wide effects of varying load ratios (dead to live)
- system wide effects of varying fuel loads, synthetic combustible fuel loads
- time to collapse without water suppression.

6.2 New Construction Systems

(or types that are not well represented by the standard fire test)

- Continuous floors and beam systems
- panel floors
- precast and prestressed hollow core concrete floors, cast in place post tensioned floors,
- open web steel joists, non compact sections, etc.

K. H. Almand, *Structural Fire Resistance Experimental Research*,
SpringerBriefs in Fire, DOI: 10.1007/978-1-4614-8112-6_6,
© Fire Protection Research Foundation 2012

- steel and polyurethane sandwich panels
- truss behavior
- large scale timber structures.

6.3 New Construction Materials

(which result in structural failure mechanisms which are different than those that serve as the basis for calculation methods)

- High strength concrete elements (spalling)
- composite wood elements (delamination), etc.
- aerated autoclave concrete
- new firestopping materials.

6.4 Validation of Full Scale Connection Behavior

- Expansion forces
- torsional failure modes
- restrained and unrestrained behavior.

Chapter 7
Research Needs Identified at Workshop

The final resource for input for research needs and priorities was a one-day workshop held on September 8, 2011 at the NIST facilities in Gaithersburg, MD. Invited participants included all those who provided input to the literature review and community input steps described above. Appendix A is the full report of that workshop, including attendance, agenda and full input from all participants. Workshop participants formed three groups to discuss steel, concrete and general large system interaction research needs and a rich description of this discussion is included in the Appendix. Technology transfer and collaborative mechanisms for research were also discussed. Priorities resulting from the group discussions are as follows:

7.1 Large System Interaction Group Priorities

- Benchmarking studies during commissioning to establish the reliability of the testing apparatus and data acquisition.
- Development of robust and reliable instrumentation to provide high resolution data necessary to validate modeling techniques.
- Evaluation of performance of existing construction methods in 3-D realistic fires.
- Develop high quality datasets for validation of analysis methods.
- Fire-structure interaction studies with particular emphasis on how the structural response can impact the fire scenario and the resulting fire effects on the structure.
- Develop prototype research concepts that fully illustrate the capabilities of the facility for industry, government and the public.
- Continued development of a comprehensive performance-based framework that is necessary to fully realize the potential innovations of advanced structural fire testing and simulation.

K. H. Almand, *Structural Fire Resistance Experimental Research*,
SpringerBriefs in Fire, DOI: 10.1007/978-1-4614-8112-6_7,
© Fire Protection Research Foundation 2012

7.2 Concrete Group Priorities

- Validation of instrumentation prior to initiation of testing.
- Explore the resiliency of concrete structural systems under extreme loading.
- Explore the concept of sustainable structural design for fire in concrete structures—i.e., post fire issues.
- Study fire performance of repaired structures.
- Evaluate the actual performance of structural elements and structural systems against the benchmark of prescriptive measures currently in building codes.
- Develop and validate methods to scale test results.
- Investigate the mechanisms for fire induced progressive collapse.

7.3 Steel Group Priorities

- Investigate multi hazard performance including consideration of community impacts (i.e., collapse, contiguous structures, sustainability).
- Development and validation of an improved prescriptive/simplified design approach that is not based on E119.
- Develop and provide a clearinghouse for thermal and mechanical properties of materials at elevated temperatures.
- Conduct a thorough literature review before starting—do not reinvent the wheel.
- Conduct a comprehensive study on the performance of floor systems uniquely suited to this facility.
- Quantify/evaluate the level of safety required/embodied in current prescriptive system.
- Develop standard fire loads, design fires.

7.4 Collaboration Mechanisms and Reaching Out to Industry

Workshop participants provided suggestions and recommendations for mechanisms to collaborate with other institutions currently engaged in structural fire engineering research as well as mechanisms for industry support. They included:

- Funded guest researchers to support experiments—knowledge transfer.
- Consortia (CRADA)—in-kind support (materials, construction).
- Industry funded research by contract.
- NSF support.

Participants suggested that more work is needed to translate the unique capability of the facility into ideas/visions of what it could actually do/problems it could address.

They suggested a regular technology transfer forum—such as an annual conference—to present results and refine and refresh collaborative opportunities and research priorities.

7.5 Overall Priorities: All Groups

In a workshop plenary session, the following seven research topics were identified as of highest priority for the facility.

- Before research begins, carry out a comprehensive assessment of the validity of measurements from the unique instrumentation systems in this facility.
- A study on scaling including validating scaling methods could extend the application of results from other laboratories.
- Resiliency—this issue is of current concern and can uniquely be validated in the NIST facility.
- The facility provides a unique opportunity to investigate 3D structural performance to develop data for model validation.
- Explore the interaction of fire effects on a structure in turn influencing fire development in the structure.
- Conduct a study on the performance of floor systems uniquely suited to this facility.
- Quantify/evaluate the level of safety required/embodied in the current prescriptive system.

Chapter 8
Synthesis of Identified Priorities for Utilization of the NFRL

Based on the input from the activities above, the following research priorities are identified for the NFRL.

8.1 Validation of Instrumentation

Because of the unique nature of the facility, there is a priority need for a thorough benchmarking and validation process of the measurement systems used in the facility. It also presents the opportunity to evaluate new and developing sensor technologies.

8.2 Fire Exposures

Although the primary focus of the NFRL is structural behavior in fire, most input included a focus on the unique ability of the facility to explore the interaction of fire exposure and structure as follows:

- Non standard fire exposures to include rapid rate of rise impacts on structural fire proofing of large loaded systems
- Varying location of localized fire exposure
- Multi-compartment fire exposure—open areas, simultaneous exposures on either side of vertical or horizontal compartment barriers
- Interaction of structural response (barrier integrity) and fire spread
- Large scale cooling effects
- Multi hazard effects

K. H. Almand, *Structural Fire Resistance Experimental Research*,
SpringerBriefs in Fire, DOI: 10.1007/978-1-4614-8112-6_8,
© Fire Protection Research Foundation 2012

8.3 Structural Details

A recurring theme amongst input on this topic was the need to focus on the unique capabilities of the NFRL and to tie this work to work ongoing at other facilities at a smaller scale. The following particular issues were identified:

- Connection behavior (large scale validation of smaller scale research results)
- Asymmetrical structural elements
- Localized failure effect at the large scale
- Detailing robustness (fire stopping, connections); quality issues;
- Repaired structures

8.4 Structural Systems

The unique configuration of the NFRL makes it most suited to research related to structural systems—systems whose performance can only be estimated based on fire testing of individual elements. The following particular issues were identified:

- Load ratios and redistribution effects
- New structural systems (in particular those for whom single element tests to do indicate system behavior)
- Large-scale timber structural performance
- Floor system performance

8.5 Materials Issues

There were three major categories of research needs identified in this area. The first relates to the thermal performance of fire proofing materials when applied to a loaded structural system (deformation effects). The second is the determination of equivalent structural properties at elevated temperatures. The third which was identified in most forums was an in-depth study of the concrete spalling issue as it applies to the large scale (again deformation effects).

8.6 Data to Inform Design Methods and Codes and Standards

Information from all of the above recommended studies is designed to inform performance-based design methods for fire as well as the design standards which are based on them. In particular, the group noted the need for:

- Development of equivalent material properties for design methods
- Understanding of the ability (or not) to scale structural effects
- Understanding of the performance of structures in comparison with predictions based on prescriptive methods to quantify embodied safety factors in those methods

Chapter 9
Conclusions

The NFRL presents a unique opportunity to explore a broad range of unanswered questions regarding the performance of real structures in the fire condition, and to inform performance based design methods and standards in this field. Although input was sought from a broad range of sources in this study, the following major issues are of broad concern to the community and are recommended for consideration as priority research areas for the NFRL in support of its objective:

1. Because of the unique nature of the facility, there is a priority need for a thorough benchmarking and validation process for the measurement systems used in the facility.
2. Although the primary focus of the NFRL is structural behavior in fire, a priority for the design community is the interaction of real fire exposure and structural response—how one affects the other.
3. A focus on large scale experiments related to the many unanswered questions about composite floor system performance would have great practical import and a major impact on design methods.
4. Material (structural and fire proofing) properties under load at the large scale are a high priority need for enhancing modeling of performance in fire.
5. Understanding the embedded safety factors in our current prescriptive design methods is an important first step in moving toward a performance based design system.
6. There is a strong interest within the structural fire engineering research community in collaborating with NIST in undertaking synergistic research projects that take full advantage of the NFRL capabilities.

K. H. Almand, *Structural Fire Resistance Experimental Research*,
SpringerBriefs in Fire, DOI: 10.1007/978-1-4614-8112-6_9,
© Fire Protection Research Foundation 2012

Appendix A
Workshop Report

THE
FIRE PROTECTION
RESEARCH FOUNDATION

Structural Fire Research Needs

Workshop Agenda
NIST Engineering Laboratory
Gaithersburg, MD
11:00 a.m. – 4:00 p.m. September 8, 2011

11:00 a.m.	Welcome, Project Background, Workshop Goal	Gross
11:30	NFRL Description	Cauffman
Noon	Strawman Research Needs Agenda	Almand
12:30 p.m.	Lunch/Break out Sessions:	
	Steel Structures:	Carter/Engelhardt
	Concrete Structures:	Szoke/Kodur
	Whole Building Behavior:	Beyler/Deierlein
2:30 p.m.	Session Reports/Prioritization/Partnerships	Almand
3:30	Adjourn/Tour of NFRL Site	

K. H. Almand, *Structural Fire Resistance Experimental Research*,   51
SpringerBriefs in Fire, DOI: 10.1007/978-1-4614-8112-6,
© Fire Protection Research Foundation 2012

Structural Fire Research Needs Workshop

September 8, NIST

Participants

Farid Alfawakhiri	AISI
Roger Becker	Precast/Prestressed Concrete Institute
Craig Beyler	Hughes Associates
Luke Bisby	University of Edinburgh
Gary Broughton	Hoover Treated Wood Products
Charlie Carter	AISC
Greg Deierlein	Stanford University
Mike Engelhardt	University of Texas
Maria Garlock	Princeton University
Jennifer Goupil	Structural Engineering Institute
Steve Hill	Rolf Jensen and Associates
Dan Howell	FM Global Corporation
Morgan Hurley	SFPE
Nestor Iwankiw	Hughes Associates
Rudy Jagnandan	Isolatek International
Ann Jeffers	University of Michigan
Venkatesh Kodur	Michigan State
Manfred Korzen	BAM (Germany)
Nicholas Lang	National Concrete Masonry Association
Gilberto Mosqueda	MCEER
Hossein Mostafaei	National Research Council, Canada
Kevin Mueller	University of Notre Dame
Darlene Rini	Arup
Serdar Selamet	Princeton University
Kuma Sumathipala	American Forest and Paper Association
Steve Szoke	Portland Cement Association
Dave Thomas	Fairfax County
Amit Varma	Purdue University
Robert Wessel	Gypsum Association
Ehab Zalok	Carleton University
Kathleen Almand	FPRF
John Gross	NIST
Steve Cauffman	NIST
Anthony Hamins	NIST
Terry McAllister	NIST

Individual Break Out Group Discussions

Large System Interaction Group

Attendees

Farid Alfawhakiri	AISI
Craig Beyler	Hughes Associates
Gary Broughton	Hoover Treated Wood Products
Greg Deierlein	Stanford University
Jennifer Goupil	Structural Engineering Institute
Dan Howell	FM Global Corporation
Morgan Hurley	SFPE
Steve Hill	RJA Group
Hossein Mostafaei	NRC Canada
Gilberto Mosqueda	MCEER
Kuma Sumathipala	American Forest and Paper Association
Dave Thomas	Fairfax County
Amit Varma	Purdue University
Robert Wessel	Gypsum Association
Ehab Zalok	Carleton University
Anthony Hamins	NIST
Terry McAllister	NIST

Large System Interaction Group

Summary by Craig Beyler and Greg Deierlein

We had a diverse breakout group who were largely well engaged. The more technically focused were engaged from the start, and the more commercially minded warmed as the meeting continued. The commercially minded made it clear that NIST has not yet communicated the actual capabilities of the facility in meaningful terms and needs to do so. The commercially focused also made it clear that any research needed to have clear commercial impact. NRCC was raised as an excellent model of industry/govt cooperation. It is a model worth studying.

The suggestions of the group were largely implicitly based upon the ability to do 3-D real scale testing with realistic fires.

There were very constructive suggestions regarding how to move from here to an effective facility. These focused on the development of instrumentation which can provide good data in this challenging environment as well as the need to build credibility in the commissioning phase by performing benchmark, essentially understood, tests which reproduce known results, hopefully with enhance instrumentation that deeps our understanding.

There was a clear sense that the ultimate goal of the facility is to allow validation of simulation models and support structural fire protection (SFP) to be done at the level of room temperature structural engineering (SE). This will require careful planning, reporting and archiving of test results to enable their use by the structural fire research and engineering communities. Several investigators embraced the hybrid testing approach as an efficient and effective use of resources that involved collaboration among laboratories and analysts.

While many focused on practical outcomes, others emphasized the need for realistic testing which is fundamental and exploratory, looking for phenomena that are significant but not now recognized. In fact, this is needed to assure appropriate application of models in SFP.

There were suggestions that involved testing of the most common assemblies and suggestions that focused novel construction methods, including the use of multiple materials within one structure. The testing involving common assemblies can be used to assess if our current construction methods and materials (designed in accordance with current building codes) are better or worse than we assume when used in real scale applications with realistic fire exposures. Benchmarking the performance of current design and construction methods was cited as a high priority and fruitful area for testing. Such studies have the potential to create an environment of innovation to cure problems or to make the most of excellent performance. In short, such testing can stir the pot, making innovation possible and an economic necessity.

Multi-hazard studies and post-fire structural assessments (residual capacity of various structures and materials) were seen as cost effective ways to get the most out of a test setup. In addition, there was an interest in the effect of earthquakes on structures and the fire performance of these degraded structures. Post-fire assessment is of economic importance after fire losses and can allow the development of better post-fire structural repairs and remediation methods. Repairs can be made in situ and the structural performance assessed in situ.

There was definite interest in studying realistic fire growth as well as studying the effect of structural deformation on fire barriers and other aspects of fire development. This would provide important knowledge to validate design and simulation methods, and it could lead to improved methods for fire barrier construction.

The priority list of tests and other activities included:

- Benchmarking studies during commissioning to establish the reliability of the testing apparatus and data acquisition.
- Development of robust and reliable instrumentation to provide high resolution data necessary to validate modeling techniques.

- Evaluation of performance of existing construction methods in 3-D realistic fires
- Develop high quality datasets for validation of analysis methods
- Fire-structure interaction studies with particular emphasis on how the structural response can impact the fire scenario and the resulting fire effects on the structure.
- Develop prototype research concepts that fully illustrate the capabilities of the facility for industry, government and the public.
- Continued development of a comprehensive performance-based framework that is necessary to fully realize the potential innovations of advanced structural fire testing and simulation.

Other topics that were suggested for investigation but that did not rise as high on the priority list include:

- Thermal expansion effects, whose mitigation may require thermal insulation to reduce temperatures to significantly lower values than the 550 °C implied by typical prescriptive requirements.
- Influence of gravity design load combinations on fire endurance, including reconciliation of inconsistent requirements from different building codes and standards.
- Development of performance-based approaches that rigorously characterize and propagate the effect of uncertainties through the performance assessment.
- Investigate how the Department of Energy requirements for green (thermally efficient) buildings are affecting fire performance, e.g., through the requirement for more thermal insulation and tighter enclosures that will tend to make fires more ventilation controlled.

Concrete Group

Attendees

Roger Becker	Precast/Prestressed Concrete Institute
Luke Bisby	University of Edinburgh
Venkatesh Kodur	Michigan State
Manfred Korzen	BAM (Germany)
Nicholas Lang	National Concrete Masonry Association
Kevin Mueller	University of Notre Dame
Steve Szoke	Portland Cement Association
Steve Cauffman	NIST

Concrete Group Notes

By Steve Szoke and Venkatesh Kodur

The concrete group discussed the need for research on the following topics:

- Validate measurement techniques—for pore pressures, rotation at the connection, strain/displacement, etc
- Fire induced progressive collapse—how to identify incipient collapse?
- Life line structures—bridges, tunnels
- Concrete filled tubes—shear, load transfer
- Sustainable design for concrete—eg., reduced cover requirement based on fire experiments
- Resilience—reuse of structural frame following fire
- Couple with blast loading
- Protection of the core inside CMU (concrete masonry unit) structural walls
- Post earthquake fire for buildings and infrastructure
- Structural material properties of concrete (fire, structural loading)
- Measurement techniques
- Performance of strengthened concrete structures (e.g. FRP reinforcement)
- Evaluation of real performance of prescriptive designs
- Experimental data to validate models of innovative connections under fire/structural loads
- Evaluate effects of scale
- Compare E119 results to real fire conditions
- Innovative product testing
- Overall strategy for implementing PBD
- Interaction of steel and concrete structural systems
- Mega floor behavior

Steel Group

Attendees

Ann Jeffers	University of Michigan
Charlie Carter	AISC
Darlene Dini	Arup
Mike Engelhardt	University of Texas
Maria Garlock	Princeton University
Nestor Iwankiw	Hughes Associates
Rudy Jagnandan	Isolatek International
Serdar Selamet	Princeton University
John Gross	NIST

Steel Group Notes

By Charlie Carter and Mike Engelhardt

Mike Engelhardt started the discussion by reviewing a concept of a major project on steel floor systems which would address the following:

- Mechanisms of load resistance.

  - flexure
  - catenary action
  - membrane action

- Influence of surrounding structural elements.

  - interior vs exterior bays

- Influence of bay dimensions and aspect ratio.
- Behavior of non-rectangular bays.
- Behavior and influence of beam-to-girder and girder-to-column connections.
- Influence of degree of composite action.

  - full vs. partial vs. none

- Behavior of shear studs.
- Influence of decking type and geometry.
- Influence of slab reinforcement.
- Behavior of systems with girders and beams made of:

  - rolled wide flange sections
  - open web steel joists and joist girders
  - others (castellated sections, slim-floor sections, etc).

- Behavior of floor systems without concrete slab.
- Build on world-wide research efforts on unprotected floor systems (Cardington and others).
- Potentially significant impact on practice and competiveness of structural steel
- Contribute to performance based design methodologies.
- Can serve as basis for improvements to ASTM E119 .
- Takes advantage of unique capabilities of the NFRL.

His recommendation for the group to consider was to focus first on the system behavior in floor systems, which seem in many cases to be capable of withstanding fire effects without fire protection materials due to catenary and membrane action. He proposed that the work could result in significant economy of future construction. Discussion ensued:

1. It was noted that this work might also contribute to performance-based design. Perhaps an owner will decide to add fire protection anyway to limit fire damage or deflection.

2. Needs include a better way to address long spans and odd configurations without penalizing usual cases that are more straightforward.
3. Current systems and approaches are viewed as safe, but the actual level of safety of a prescriptive system is not well known. We need research that will help us predict this level of safety better, and perhaps also permit designs to be economized from what is currently done in prescriptive approaches.
4. Fire loads and design fires need to be defined.
5. What about unsymmetrical fire exposures?
6. Boundary conditions for fire loads need to be defined.
7. Column performance and column bracing performance in a fire needs to be studied.
8. A clearinghouse is needed for information on detailed thermal properties of the various fire-protective materials (including those that are proprietary).
9. Need real system tests to help better understand FEM results.
10. Need to assess scale effects and how they should be treated when doing tests.
11. Work should be done to enable the EOR to provide fire engineering services, perhaps with a simplified approach that is better than the current prescriptive approach, but not necessarily as advanced as a PBD fire engineering approach.
12. Better prediction of cracking and falling off of fire protective materials is needed.
13. Durability requirements should be better developed. Right now, they simply represent the threshold of what current products provide. The values given should be based upon what bond strength is needed, not what is available from all manufacturers.
14. Multi-hazard scenarios should be investigated (blast followed by fire, earthquake followed by fire, etc.)
15. Bridge fire exposures should be studied.
16. Concrete performance in fire, especially for current products and higher-strength concrete products, is not well understood. Some products perform poorly, like explosive spalling in super-high-strength concrete. This remains of interest to define for steel focused people because of composite design and construction.
17. Research should be undertaken to properly characterize the catenary and membrane action behavior that currently is poorly captured by the restrained vs. unrestrained argument. The group believes that the issue is meaningless when one looks at the system behavior.
18. Work out a coordination process for how to collaborate with university researchers.
19. Technology transfer is needed when results are available. Do an annual conference? Prepare for seminars? Provide education in some capacity?
20. Can model provisions for fire loads be developed as a prestandard that could then be used to develop standard provisions in ASCE 7?

Individual Group Priorities

Large System Interaction

Benchmark against what we know during commissioning/credibility
   Develop data to validate models, 3-D behavior, finite element models in particular
   Explore the interaction of fire effects on structure in turn influencing the fire development of structures

Concrete

Instrumentation validation
   Resiliency under extreme loading
   Sustainable fire designs
   Fire performance of repaired structures
   Real performance against prescriptive measures
   Methods to scale test results
   Fire induced progressive collapse

Steel

Multi hazard performance including consideration of community impacts
   Development and validation of an improved prescriptive/simplified design approach that is not based on E119
   Develop and provide a clearinghouse for thermal and mechanical properties of materials at elevated temperatures
   Conduct a thorough literature review before starting—don't reinvent the wheel
   Conduct a comprehensive study on the performance of floor systems uniquely suited to this facility
   Quantify/evaluate the level of safety required/embodied in current prescriptive system
   Develop standard fire loads, design fires

Collaboration Mechanisms and Reaching Out to Industry

Funded guest researchers to support experiments—knowledge transfer
   Consortia (CRADA)—in-kind support (materials, construction)
   Industry funded research by contract
   NSF support
   Need to develop a means to translate the capability of the facility into ideas/visions of what it could actually do/problems it could address

Technology Transfer

An annual conference, partner with MHMC or others to develop model code provisions etc

Overall Priorities—All Groups

1. Before research begins, carry out a comprehensive assessment of the validity of measurements from the unique instrumentation systems in this facility
2. A study on scaling, validating scaling methods, could extend the application of results from other laboratories
3. Resiliency—this issue is of current concern and can uniquely be validated in the NIST facility
4. 3-D structural performance model validation data
5. Explore the interaction of fire effects on structure in turn influencing fire development in structures
6. Conduct a study on the performance of floor systems uniquely suited to this facility
7. Quantify/evaluate the level of safety required/embodied in the current prescriptive system

References

AISC (2005) ANSI/AISC 341-05—Seismic provisions for structural steel buildings. American Institute of Steel Construction, Chicago, p 334

Almand, K., Phan, L., McAllister, T., Starnes, M., & Gross, J. (2004). *SFPE workshop for development of a national R&D roadmap for structural fire safety design and retrofit of structures, NISTIR 7133*. Gaithersburg: National Institute of Standards and Technology. 188.

Anderberg Y (1986) Modelling steel behaviour, presented at the International conference on design of structures against fire, Aston University, Birmingham, April 15–16, pp 5

Annerel, E., Lu, L., & Taerwe, L. (2011). *Punching shear tests on flat concrete slabs at high temperatures, Proceedings of the 2nd international RILEM workshop on concrete spalling due to fire exposure, 5–7 October* (pp. 125–131). The Netherlands: Delft.

Fire A (2005) Fire resistance of concrete enclosure—work package 2: spalling categories, elaborated for the nuclear safety directorate of the health and safety executive, London, p 36

ASTM (2011) ASTM E119-11a: standard methods of fire test of building construction and materials, American society for testing and materials, West Conshohocken, p 34

Bailey, C. G. (2002). Holistic behaviour of concrete buildings in fire. *Proceedings of the institution of civil engineers, structures and buildings, 152*(3), 199–212.

Bailey, C. G., Lennon, T., & Moore, D. B. (1999). The behaviour of full-scale steel framed buildings subjected to compartment fires. *The Struct Eng, 77*(8), 15–21.

Bailey, C. G., & Toh, W. S. (2007). Behaviour of concrete floor slabs at ambient and elevated temperature. *Fire Safety J, 42*(6–7), 425–436.

Bailey, C. G., & Lennon, T. (2008). Full scale fire tests on hollowcore slabs. *The Struct Eng, 86*(6), 33–39.

Bailey, C. G., & Khoury, G. A. (2011). *Performance of concrete structures in fire—an in-depth publication on the behaviour of concrete in fire*. Reading: Ruscombe Printing Ltd. 203.

Bamonte, P., Felicetti, R., & Gambarova, P. G. (2009). *Punching shear in fire-damaged reinforced concrete slabs, ACI special publication 265* (pp. 345–366). Farmington Hills: American Concrete Institute.

Beitel J, Iwankiw N (2008) Analysis of needs and existing capabilities for full-scale fire resistance testing, NIST GCR 02-843-1 (Revision), National institutes for standards and technology, October, p 98

Beyler, C., Beitel, J., Iwankiw, N., & Lattimer, B. (2007). *Fire resistance testing for performance-based fire design of buildings, NIST GCR 07-910*. Gaithersburg: National Institute of Standards and Technology. 154.

Bisby, L. A., & Kodur, V. K. R. (2007). Evaluating the fire endurance of concrete slabs reinforced with FRP bars: Considerations for a holistic approach. *Composites Part B: Eng, 38*(5–6), 547–558.

K. H. Almand, *Structural Fire Resistance Experimental Research*,
SpringerBriefs in Fire, DOI: 10.1007/978-1-4614-8112-6,
© Fire Protection Research Foundation 2012

BRE (2011) Fire test facilities. Available at http://www.bre.co.uk/page.jsp?id=417. Accessed 7th December 2011

Steel, British. (1999). *The behaviour of multi-storey steel frame buildings in fire*. Rotherham: British Steel. 82 pp.

CCAA (2010) Fire safety of concrete buildings, Cement Concrete and Aggregates Australia—CCAA T61, July, p 37

CEN (2002) BS EN 1991-1-2:2002—Eurocode 1: Actions on structures—Part 1–2: general actions—Actions on structures exposed to fire, European Committee for Standardization, Brussels, p 62

CEN (2004) BS EN 1992-1-2:2004—Eurocode 2: Design of concrete structures—Part 1.2: general rules—Structural fire design, European Committee for Standardization, Brussels, 100

Chlouba, J., & Wald, F. (2009). Connection temperatures during the Mokrsko fire test. *Acta Polytechnica, 49*(1), 76–81.

Chlouba, J., Wald, F., & Sokol, Z. (2009). Temperature of connections during fire on steel framed building. *Int J Steel Struct, 9*(1), 47–55.

Chung, H.-Y., Lee, C.-H., Su, W.-J., & Lin, R.-Z. (2010). Application of fire-resistant steel to beam-to-column moment connections at elevated temperatures. *J Construct Steel Res, 66*, 289–303.

Corus (2006) Fire resistance of steel-framed buildings, Corus Construction and Industrial, p 40

CSTB (2011) Vulcain. Available at http://www.cstb.fr/le-cstb/equipements/feu/vulcain.html. Accessed 1st December 2011.

De Feijter, M. P., & Breunese, M. P. (2007). *2007-Efectis-R0894(E)—Investigation of fire in the Lloydstraat car park, Rotterdam*. The Netherlands: Efectis. 50.

Ding, J., & Wang, Y. C. (2007). Experimental study of structural fire behaviour of steel beam to concrete filled tubular column assemblies with different types of joints. *Eng Struct, 29*(12), 3485–3502.

Dong, Y., & Prasad, K. (2009). Thermal and structural response of a two-story, two bay composite steel frame under fire loading. *Proc Combustion Inst, 33*, 2543–2550.

Dong, Y., & Prasad, K. (2009). Experimental study on the behavior of full-scale composite steel frames under furnace loading. *J Struct Eng, 135*(10), 1278–1289.

FEMA (2000a) FEMA-350—Recommended seismis design criteria for new steel moment-frame buildings, Federal emergency management agency, p 224

FEMA (2000b) FEMA-3555D—State of the art report on connection performance, prepared for the SAC Joint Venture Partnership, Federal emergency management agency, 305

Firehouse.com (2004) Seven swiss firefighters die in collapsed parking garage, available at http://www.firehouse.com/news/lodd/seven-swiss-firefighters-die-collapsed-parking-garage. Accessed on 6th December 2011.

Global, F. M. (2009). *The FM Global Research Campus*. USA: FM Global. 20.

Gillie, M. (2009). Analysis of heated structures: nature and modelling bench marks. *Fire Safety J, 44*(5), 673–680.

Grosshandler WL (2002) Fire resistance determination and performance prediction research needs workshop: proceedings, NISTIR 6890, National institute of standards and technology, 162

Gorsshandler, W. L. (2003). The international FORUM of fire research directors: a position paper on evaluation of structural fire resistance. *Fire Safety J, 38*, 645–650.

Han, L. H., Wang, W. H., & Yu, H.-X. (2010). *Performance of RC beam to concrete filled steel tubular (CFST) column frames subjected to fire, Proceedings of the 6th international conference on structures in fire (SiF'10), June 02–04* (pp. 358–365). East Lansing, Michigan: Michigan State University.

Huang, S. S., Burgess, I., & Davidson, B. (2011). *A structural fire engineering prediction for the Veselí fire tests, 2011, Proceedings of the international conference on applications of structural fire engineering, April 29* (pp. 411–416). Prague: Czech Republic.

ISO (1999) ISO 834-1:1999: fire resistance tests - elements of building construction -Part 1: general requirements. International organization for standardization, Geneva, Switzerland, p 25

Kelly, F., & Purkiss, J. (2008). Reinforced concrete structures in fire: a review of current Rules. *The Struct Eng, 86*(19), 33–39.

Kodur VKR, Garlock MEM, Iwankiw N (2007) Structures in fire: state of the art, research and training needs. NIST Report GCR 07-915. National institute of standards and technology, Gaithersburg, p 63

Kodur VKR, Garlock MEM, Iwankiw N (2011) Structures in fire: state of the art, research and training needs. Fire technology (in press)

Kordina K (1997) Ueber das brandverhaltem punkthestuetzter stahlbetonplatten (Investigations on the behaviour of flat slabs under fire). DAfsib H. 479. 106S.

Korzen M, Rodrigues J, Correia A (2010) Composite columns made of partially encased steel sections subjected to fire. Proceedings of the 6th international conference on structures in fire (SiF'10), June 02–04, Michigan State University, East Lansing, Michigan, pp 341–348

Law M (1981) Designing fire safety for steel—recent work, Proceedings of the ASCE spring convention, American Society of Civil Engineers, New York, 11–15 May, p 16

Lennon T, Bullock MJ, Enjily V (2000) The fire resistance of medium-rise timber frame buildings. World Conference on Timber Engineering, Whistler Resort, British Columbia, Canada, July 31–August 3, p 10

Li-tang G, Li X, Chen L, Yuan A (2008) Experimental investigation of the behaviours of unbonded prestressed concrete continuos slabs after fire, Concrete, p 220 (Chinese)

Lv J, Dong Y, Yang Z, Sun J (2011) Experimental and analytical studies on performance of edge beams of steel framed building subjected to fire, J Build Struct (Chinese) Sept

Mostafaei H (2011) Hybrid fire testing for performance evaluation of structures in fire—Part 1: Methodology, Research Report No. RR-316, NRC institute for research in construction, Aug 26, p 22

New York Times (1997) Philadelphia to Raze Site of High-Rise Fire. Available at http://www.nyt imes.com/1997/11/14/us/philadelphia-to-raze-site-of-high-rise-fire.html, November 14, 1997, accessed 7th December 2011

NIST (2011) Project: national fire research laboratory commissioning and operations. Available at http://www.nist.gov/el/fire_research/nfrl/project_nfrl.cfm, accessed 5th Dec 2011

NRCC (2011) Burn hall—three storey exterior wall test facility. Available at: http://www.nrc-cnrc.gc.ca/eng/facilities/irc/burn-hall.html, accessed 9th December 2011

Proe D, Thomas I (2010) Planning for a large-scale fire test on a composite steel-frame floor system. Proceedings of the 6th international conference on structures in fire (SiF'10), June 02-04, Michigan State University, East Lansing, Michigan, pp 390–397

Peng, W., Hu, L.-H., Yang, R.-X., Lv, Q.-F., Tang, F., Xu, Y., et al. (2011). Full scale test on fire spread and control of wooden buildings. *Procedia Eng, 11*, 355–359.

Ring, T., Zeiml, M., & Lackner, R. (2011). *Large scale fire tests on concrete design and results, presented at the 18th inter-institute seminar for young researchers (IIS18), Sep 23–25.* Hungary: Budapest.

Robert F, Rimlinger S, Collet E, Collignon C (2009) PROMETHEE, the Innovative fire resistance testing centre for structures. International conference applications of structural fire engineering, Feb 19–21, Prague, Czech Republic, Annex 14

Santiago, A., da Silva, L. S., Vaz, G., Vila Real, P., & Lopez, A. G. (2008). Experimental investigation of the behaviour of a steel sub-frame under a natural fire. *Steel and Composite Struct, 8*(3), 000–000.

Sharma UK, Bhargava P, Singh B, Singh Y, Kumar V, Kamath P, Usmani A, Torero J, Gillie M, Pankaj P, May I, Zhang J (2012) Full scale testing of a damaged RC frame in fire. Proceedings of the institution of civil engineers, structures and buildings, special issue on Structures in Fire (in press)

SINTEF (2011) Large test hall, available at http://www.sintef.no/home/Building-and-Infrastructure/SINTEF-NBL-as/About-SINTEF-NBL/Laboratories/Large-test-hall/, accessed 9th December 2011.

SwRI (2011) Fire testing services, Available at: http://www.swri.org/4org/d01/fire/fireres/home.htm, accessed 5th December 2011

Stadler M, Mensinger M, Schaumann P, Sothmann J (2011) Munich fire tests on membrane action of composite slabs in fire—test results and recent findings. Proceedings of the international conference on applications of structural fire engineering, Apr 29, Prague, Czech Republic, pp 177–182

Terrasi GP (2007) Prefabricated thin-walled structural elements made from HPC prestressed with pultruded carbon wires, Proceedings of the 8th international symposium on fiber reinforced polymer reinforcement for concrete structures (FRPRCS-8), University of Patras, Greece, July 16–18, p 10

TFRI (2011) I hold China-US fire protection technology and fire codes academic exchange activities, available at http://www.tfri.com.cn/manage/html/1356.html, accessed 9th Dec 2011

Troxell, G. E. (1959). Fire resistance of a prestressed concrete floor panel. *J Am Concrete Inst, 56*(8), 97–105.

Van Acker, A. (2003). Shear resistance of prestressed hollow core floors exposed to fire. *Struct Concrete, 4*(2), 65–74.

Van Herberghen P, Van Damme M (1983) Fire resistance of post-tensioned continuous flat floor slabs with unbonded tendons, FIP Notes, pp 3–11

Vassart O, Bailey C, Hawes M, Nadjai A, Simms W, Zhao B, Gernay T, Franssen JM (2010) Large-scale fire test of unprotected cellular beam acting in membrane action. Proceedings of the 6th international conference on structures in fire (SiF'10), June 02–04, Michigan State University, East Lansing, Michigan, pp 398–406

Vassart O, Zhao (2011) FRACOF engineering background. Report developed for the project Leonardo Da Vinci: fire resistance assessment of partially protected composite floors, p 132

Victoria University (2011) Fire test facilities. Available at http://www.vu.edu.au/centre-for-environmental-safety-and-risk-engineering-cesare/fire-test-facilities. Accessed 7th December 2011

Wald, F., Simões da Silva, L., Moore, D. B., Lennon, T., Chaldna, M., Santiago, M., et al. (2006). Experimental behaviour of a steel structure under natural fire. *Fire Safatey J, 41*(7), 509–522.

Wald, F. (2010). *Fire Test on an Administrative Building in Mokrsko.* Printing house Česká technika: Czech Technical University in Prague. 152.

Wald F (2011) Large scale fire tests, COST action TU0904: integrated fire engineering and response working group, Prague, Czech Republic, Apr 30, p 5

Wald, F., Jána, T., & Horová, K. (2011). *Design of joints to composite columns for improved fire robustness - To demonstration fire tests.* Printing house Česká technika: Czech Technical University in Prague. 26.

Wang, Y. (2002). *Steel and composite structures: behaviour and design for fire safety.* London: Spon Press. 332.

Wong, Y. L., & Ng, Y. W. (2011). *Technical seminar—effects of water quenching on reinforced concrete structures under fire, presented at the institution of fire engineers (Hong Kong Branch), Aug 23.* Hong Kong: Kowloon Tong Fire Station.

Wong, S. Y., Burgess, I. W., Plank, R. J., & Atkinson, G. A. (1999). The response of industrial portal frames to fires. *Acta Polytechnica, 39*(5), 169–182.

Yu, H., Burgess, I., Davison, J., & Plank, R. (2011). Experimental and numerical investigations of the behavior of flush end plate connections at elevated temperatures. *J Struct Eng, ASCE, 137*(80), 80–87.

Yu, H., Burgess, I. W., Davison, J. B., & Plank, R. J. (2009). Experimental investigation of the behaviour of fin plate connections in fire. *J Construct Steel Res, 65*(3), 723–736.

Yuan, Z., Tan, K. H., & Ting, S. K. (2011). Testing of composite steel top-and-seat-and-web angle joints at ambient and elevated temperatures: part 2—elevated-temperature tests. *Eng Struct, 31*(9), 2093–2109.

Zhao, J. C., & Shen, Z. Y. (1999). Experimental studies of the behaviour of unprotected steel frames in fire. *J Construct Steel Res, 50*, 137–150.

Zheng, W. Z., Hou, X. M., Shi, D. S., & Xu, M. X. (2010). Experimental study on concrete spalling in prestressed slabs subjected to fire. *Fire Safety J, 45*, 283–297.